Author :- GAURAV BHADANI

DIRECTOR OF BHADANI QUANTITY SURVEYING

AND TRAINING PVT LTD

https://www.constructionmanagementinstitute.com/

www.billingengineer.com

www.bhadaniinternaional.com

www.estimationandcosting.com

DISCLAIMER

Although the publisher and the author have made every effort to ensure that the information in this book was correct at press time and while this publication is designed to provide accurate information in regard to the subject matter covered, the publisher and the author assume no responsibility for errors, inaccuracies, omissions, or any other inconsistencies herein and hereby disclaim any liability to any party for any loss, damage, or disruption caused by errors or omissions, whether such errors or omissions result from negligence, accident, or any other cause.

This book is not intended for use as a source of legal, business, accounting or financial advice. All readers are advised to seek services of competent professionals in legal, business, accounting and finance fields.

You are encouraged to print this book for easy reading.

TABLE OF CONTENTS

SECTION A: GENERAL QUALITY OF MATERIALS AND WORKMANSHIP

SECTION B: EXCAVATION AND EARTHWORK

SECTION C: CONCRETE WORK

SECTION D: BLOCKWORK, BRICKWORK AND MASONRY

SECTION E: ROOFING

SECTION F: CARPENTARY AD JOINERY

SECTION G: STRUCTURAL STEELWORK

SECTION H: METALWORK

SECTION I: FINISHES

SECTION J: GLAZING SECTION

SECTION K: PAINTING AND DECORATING SECTION

SECTION L: ELECTRICAL WORKS SECTION

SECTION M: MECHANICAL WORKS

SECTION N: EXTERNAL WORKS

SECTION A

MATERIAL AND WORKMANSHIP QUALITY

A1: Materials and workmanship standards are generally those issued by the British Standards Organisation (BS) or the American Society for Testing and Materials (ASTM).

A2: All materials used in the project must be new and meet the specified standards. All materials samples will be subjected to testing as directed by the Project Manager from time to time, and only those materials that pass the test will be used in the works. Any material that is damaged or faulty may be rejected by the Project Manager, and such material must be removed from the site immediately.

A3: Workmanship is to be of the highest possible standard, and where a Standard Code of Practice exists and is applicable to any portion of the Works, the Contractor shall allow for compliance with the recommended practise unless it would directly conflict with requirements stated elsewhere in these contract documents.

A4: The Contractor must provide all necessary certificates demonstrating that the materials used are in compliance with the specifications.

A5: Proprietary materials must be handled, stored, and used in strict accordance with the manufacturer's instructions and recommendations, and processes must be followed to the letter.

A6 Materials of proprietary manufacture specified hereafter may be substituted with materials of a different manufacture if such substitutes are in all respects equal to the original specification and the Project Manager's prior approval is obtained for all substitutions made.

A7: Instruments, machines, labour, and materials that are normally required for inspecting, measuring, and testing any work, as well as samples of materials for testing the quality and weight of any material used before incorporation in the works, shall be provided to the Project Manager.

Samples and Tests

A8: At the Contractor's expense, samples of all materials to be used in the works must be provided.

A9: The Contractor is responsible for all costs associated with testing.

A10: The Project Manager and any other person he authorises shall have unrestricted access to the works, including all workshops and locations where services are provided or where materials, manufactured articles, or machinery for the works are obtained.

Inspection of the Work

A11: Without the approval of the Project Manager or his representative, no work shall be covered up or hidden.

A12: The Contractor shall give the Project Manager or his representative every opportunity to inspect and measure any work that is about to be covered up or hidden.

A13: Whenever work is ready or about to be ready for examination, the Contractor must give the Project Manager or his representative adequate notice.

Descriptions that contradict each other

A14: If the detailed descriptions of any measured item or group of measured items in the Bills of Quantities conflict with the general specification clauses in the Trade Preambles, the detailed description of the measured work takes precedence.

Tolerances are a type of tolerance. A15: When a finish (e.g. rendering) is required on a structural member, special care must be taken in the structural member's construction with regard to its relationship with adjoining structural members, so that no irregularities appear on the finished surface to indicate variation in the structural members covered by that finish.

Protection A16: The Contractor is responsible for covering up and protecting all new work from injury from any source, as well as supplying all temporary doors, window protection, and any other necessary protection for all works completed, whether by himself, special tradesmen, or subcontractors. The Contractor is responsible for repairing any damaged work at his own expense.

A17: All projecting sills, mouldings, concrete steps, and the like must be protected from damage by rough timber causing security fixed during the course of the work.

A18: During the performance of the Contract, all joinery must be protected from damage from any source.

A19: At completion, the Works must be left perfect to the satisfaction of the Project Manager, and before handing over possession, the Contractor must ensure that all doors, sashes, and other components operate smoothly and make any necessary adjustments.

All Sections are subject to the specifications.

A20: The specifications for each section shall apply equally to all work of a similar nature carried out in other sections (for example, the preambles for excavation and earthwork apply equally to external works.)

A21: In these Bills of Quantities, the abbreviations BS and CP refer to the relevant British Standard Specification and British Standard Code of Practice, as published by the British Standards Institution and current at the time of tender.

A22: A hyphen between two dimensions in a description, such as 150mm-230mm, is assumed to mean "over 150mm but not exceeding 230mm," etc.

A23: In terms of metric units, the following standard abbreviations are used throughout:

mm - mm

cm stands for centimetre

m stands for metre.

m2 stands for square metre.

m3 stands for cubic metre.

N stands for Newton.

Mega Newton (MN) is a state in the United States.

Kilogramme (kg) is a unit of measurement for weight.

Number (nr)

Writing Short A24: When an item is written short, it is understood to mean that it is "extra over" the preceding item.

Prices for the contractor will be all-inclusive.

A25: The rates inserted by the Contractor in the Bills of Quantities must include all obligations contained in these documents, whether explicitly detailed or implied. If the Contractor has any questions about the scope of his obligations under any of these documents, he should seek clarification from the Quantity Surveyor. The submission of a tender by a Contractor implies that there will be no questions about the scope of his obligations under the Contract.

Bills of Quantities Prices for PCs

A26: Some Bills of Quantities descriptions include a PC Price, such as PC N3, 000/.m2. When pricing these items, the Contractor should use the P.C price as the cost of the material delivered to the job site. He must include in his rates the costs of taking delivery, unloading, handling, wastes, stacking or storing until needed, incorporating into the works, and meeting all other obligations. When a location other than the site of the works is specified, the Contractor must also allow for delivery and transportation of materials to the site at that location.

SECTION B EXCAVATION AND EARTHWORK

General

B1: Excavation must be done in accordance with the lines, levels, widths, and depths shown on the drawings or as directed by the Architect.

B2: When the excavations are ready to receive the concrete foundations, the Contractor must notify the Architect, who must approve the excavations before the concrete is poured. If the Architect requires it, any concrete or other work put in place prior to this will have to be removed at the Contractor's expense.

B3: As directed by the Architect, all black earth excavated shall be deposited in spoil heaps for re-use.

B4: No sand or ballast from the job site is to be used in the project.

B5: The Contractor is responsible for securing the sides of all excavations with shuttering, shoring, sheeting, timbering, or other means to protect the workers and prevent damage to adjacent services and structures.

B6: If the Contractor excavates to a depth or width greater than shown on the drawings or as instructed by the Architect, he must fill in the excavation with concrete as described for foundations at his own expense.

B7: All excavations must be kept free of water and mud at all times, from whatever source, by pumping and/or bailing as needed.

B8: If rock is discovered during excavation, it must be reported to the Architect.

B9: The term "rock" refers to hard stone that, in the opinion of the Architect, can only be removed by barring, wedging, drilling, splitting, or blasting. If the Contractor fails to notify the Architect prior to beginning work in rock, the excavation will be considered normal ground and valued accordingly.

B10: Rock blasting must be done with the Architect's permission, and all safety precautions, such as explosives transportation and storage, signage, and warning signals, must be followed in accordance with local laws and regulations.

B11: Before concreting, the Architect must inspect and approve the excavation's bottom. Before the excavations begin, he must be given at least 24 hours' notice. Any work done prior to inspection must be removed and redone according to the Architect's instructions if necessary.

Stuffing

B12: Wherever possible, suitable excavation material should be used for site filling, unless the Architect specifies or directs otherwise.

B13: Approved suitable materials from borrow areas approved by the Architect may be used if suitable materials are not available from excavations.

B14: The Contractor shall not fill in any work until the Architect has approved it.

B15A: Filling around foundations and other structures should be 200mm thick, with layers well rammed and consolidated, with water used to aid consolidation if necessary.

B15B: All excavated surplus materials must be deposited, spread, and levelled where directed, or removed from site as needed, with the Contractor responsible for locating a deposit and paying all fees and charges that may be imposed.

Compaction of Fills

B16: The standard proctor test, test Nos 12 and 14, shall be used to determine the compaction achieved in fills in accordance with BS 1377.

B17: To achieve the required compaction, the moisture content of the fill material must be adjusted.

A minimum relative compaction of 100% of the maximum dry density must be achieved. Any material that does not achieve this figure after repeated compaction must be removed and replaced with suitable material.

B18: All fill material, whether in place or not, that does not comply with the specification in the opinion of the Architect shall be removed and replaced as directed.

B19: The frequency of compaction tests must be such that each layer is satisfactorily compacted before moving on to the net layer with a maximum thickness of 200mm. The tests will be conducted in accordance with the Architect's instructions and under his supervision.

Components / Materials

B20 Approved filling' must be made of soft laterite or other approved materials by the Architect. It should be laid out in 300mm layers, watered, and consolidated. If excavation materials are approved for use in filling, the Contractor will agree to a reduction in the price quoted in the Bills of Quantities.

Treatment for termites / Anti Termite Treatment

B21: During excavation, the Contractor must remove all ants' nests and vermin.

Bills of Quantities Rates

B22: Unless otherwise stated, measured items of excavation shall be deemed to be no more than 1.50m deep, and rates for all excavation work shall include the following:-

1. Excavating in any material encountered, except old foundations and 'rock' as defined, as well as grubbing up and removing any drains, pipes, roots, and other obstructions.

2. Temporary sheet pilling, planking, and strutting, special shoring, re-strutting, and re-shoring, and/or any other method of supporting and maintaining the excavation faces

3. Bulk increases after excavation (pre-excavation quantities are measured for excavation and disposal).

4. Compacting the ground surface and levelling the grading to falls and cambers.

5. Trimming cuttings and embankment sides to make them slope.

6. Pumping or baling the excavations to keep storm or percolating water out. (The Contractor is responsible for determining the water table level, as well as the high and low water levels, and making appropriate provisions.)

7. Excavation to provide additional working space

8. Moving or double-handling spoils on the job site.

B23: Rates for approved filling, hardcore, and laterite shall include all necessary hand packing, ramming and levelling of the ground beneath, rolling or ramming to the surface of each layer, blinding with fine particles to the finished surface, finishing to falls or cambers as required, and loss of bulk due to consolidation. (After consolidation, the thicknesses and volumes are given.)

B24: The entire excavation work (including external works) must be carried out in order to achieve the most cost-effective balancing of spoil; for example, if surplus spoil results from the excavation of one portion of the work, that portion must be excavated in a timely manner so that the surplus can be re-used elsewhere in places where 'filling' is required. Similarly, spoil should not be carted away from the construction site if it is needed for filling.

SECTION C CONCRETE WORK

General

Cement, fine aggregate, coarse aggregate, and water are used to make concrete. Without the Architect's permission, no other agent or ingredient may be added to the concrete. The Contractor must ensure that the use of any such approved additive has no negative impact on the finished concrete works' strength, durability, or appearance.

C2: The Architect's approval in no way absolves the Contractor of responsibility for the quality of materials and workmanship used in the finished works, as well as the strength, durability, and appearance of the concrete works.

C3: The term "Testing Authority" refers to a company nominated by the Contractor and approved by the Architect that is fully equipped to conduct the tests and checks required by this Contract.

It could be a separate company, a laboratory set up and maintained by the Contractor, or a hybrid of the two.

Work that is defective

C4: Any finished works, materials, or workmanship in any part of the works that, in the opinion of the Architect, do not comply with all relevant requirements of this contract shall be classified as defective work.

C5: All defective work shall be cut out and removed from the works, with the Architect/S.O. satisfied.

C6: The extent of the work to be removed, as well as the methods to be used in removing and replacing it, shall be in accordance with the Architect's direction.

Materials that were rejected

C7: All materials that have been damaged, contaminated, deteriorated, or do not meet the requirements of these Specifications will be rejected and removed from the site at the Contractor's expense immediately.

Conceptualization / Design

C8: The reinforced concrete and precast concrete works have been designed in general in accordance with the recommendations contained in BS 110, and unless specifically excluded or modified hereafter, the Contractor shall follow the recommendations made in this Code of Practice. The following are the working tolerances.

C9: Unreinforced concrete must meet all of the contract's relevant requirements.

Components

C10: The cement must be Portland cement that meets BS.12 standards. Resistance to sulphates and rapid hardening The standards for Portland cement are BS 4027 (Part 2) and BS 12 (Part 2), respectively.

C11: All cement must be delivered to the job site in approved bulk cement lorries, sealed bags, or sealed tins, and must be fresh and free of lumps and dampness. Cement must be used in the order in which it was delivered to the job site, and no cement older than three months may be used. The Architect must approve and produce proof of the date of manufacture in advance.

C12: Before delivery, samples of each consignment must be submitted to the Architect for approval, and additional bulk samples must be provided for testing if the Architect so requests.

If the Architect finds any sample to be unsatisfactory, the entire consignment from which it was taken will be rejected, and the Architect will demand that the cement consignment be removed from the site.

C13: Cement must be stored in a weatherproof, well-ventilated shed with a dry floor raised at least 10 cm above ground level, or in approved silos.

Compositions

C14: Admixtures must only be used if this specification specifically allows them. The concrete supplier must certify to the Architect that the proposed admixtures are compatible with each other and the concrete mix ingredients. The use of admixtures will be permitted only if they do not reduce the concrete's strength below the specified minimum compressive strength.

The ASTM C260 Specification for Air-entraining Admixtures for Concrete must be followed when using C15 air-entraining admixture. The Architect must approve MBVR (Vinsol resin) or a substitute as equal before it can be used.

C16: The water-reducing admixture shall be Type A, as defined by ASTM C494, and the admixture shall be Pozzolith 30N or a substitute accepted as equal by the Architect prior to use.

C17: To colour the concrete that protects the electrical duct banks, synthetic red oxide colourant will be used. Red pigmented concrete with at least twenty-five pounds of red oxide per cubic yard must be used.

Aggregate

C18: Aggregate must follow the B.S. 882 recommendations.

C19 In exceptional circumstances, a deviation from B.S. 882 in aggregate grading may be permitted with the Architect's prior approval.

C20 Aggregate shall be stored in approved containers on hard-surfaced, self-draining areas with sufficient dividing walls to prevent the mixing of different types of aggregate.

C21 is the code for the twenty-first century The fine aggregate size limit for structural concrete grades is 5mm, and the coarse aggregate size limit is 20mm.

Fine Aggregate

C22 Fine aggregate for reinforced concrete shall be clean, well-washed, sharp, fresh water or pit sand from an approved source, or obtained by screening crushed gravel or stone, or a combination thereof, from an approved source.

C23: Fine aggregate must be free of dust, loarn, and harmful amounts of clay, shale, alkali, and organic matter, and must have hard, uncoated grains.

C24: Fine aggregate must contain no more than 3% of material passing through a 0.1mm sieve after washing, as well as no more than 1% clay lumps or 1% shale.

C25: Fine aggregate must be well graded from coarse to fine, and sieve analysis must meet the following requirements:

Sieve Size	7mm	3 mm	1 mm	0.2 mm
Percentag Passing	100	55.75	20.50	2-15

C26: Before delivery to the site, samples from each consignment of fine aggregate must be submitted to the Architect for any additional testing and approval.

Coarse Aggregate

C27: Crushed stone, gravel, or other approved inert materials with similar characteristics, or a combination thereof, shall consist of clean, hard, sound, strong, durable uncoated particles free of harmful amounts of soft, friable, thin, elongated, or laminated pieces alkali, organic, or other deleterious matter.

C28: Coarse aggregate must contain no more than 1.5 percent of material passing through a 0.2 mm sieve and no more than 5% soft fragments after washing.

C29: Coarse aggregate must be well graded between the limiting sizes and meet the sieve analysis requirements:

Sieve Size	30 mm	15 mm	7 mm	3 mm
Percent Passing	100	10 – 85	25 – 40	0 - 5

C30: Before delivery to the site, samples from each consignment of coarse aggregate must be submitted to the Architect for any additional testing and approval.

Body of water

C31: The water used in the works must be free of all harmful substances in suspension or solution.

C32: Water supply analysis will be required prior to concreting and throughout the construction process. BS 3248 will be used for testing.

Reinforcement

C33: Mild steel and high steel must be obtained from an approved supplier and must meet all BS 4449, BS 4461, BS 4482, and BS 4483 requirements.

C34: Reinforcement must be stored on racks above ground level.

C35: Before any reinforcement is ordered, the Architect must approve samples as well as certified test data for ultimate tensile stress, yield point, stress elongation, and cold bend.

C36: Steel reinforcing bars must be kept clean and free of pitting, loose rust, mill scale, oil, grease, earth, paint, or any other material that could interfere with the bond between the concrete and the reinforcement.

C37: Steel mesh and expanded metal sheet shall be used only when specifically specified, and must be of the type shown on the plan or approved by the Architect.

C38: Welding of structural reinforcement is prohibited unless the Architect expressly authorises it in writing.

C39: Concrete spacer blocks used to ensure proper placement of reinforcement must be as small as possible and of shapes and designs that will not overturn when the concrete is poured. They must be made of the same mix proportions and strength as the adjacent concrete, or of a mortar that is not less than cement-sand 1:1.

C40: All reinforcement must be properly installed and maintained in the position shown on the drawings.

C41: Reinforcement must be straightened and bent to the dimensions specified in the bar bending schedule without injuring the reinforcing bars. Bends must be made without the use of heat, using a pin with a diameter that is at least four times the bar's diameter.

Once bent, bars cannot be straightened or bent again, and bars with cracks or splits at the bends must be rejected.

C42: Soft annealed 1.6 mm diameter wire shall be used to tie all intersecting bars together, with no wire ends protruding into the concrete face. Splices in the reinforcement must only be made where indicated on the drawings. Bars must overlap for a minimum distance proportional to the diameter of the bars but no less than 30 times the diameter, as determined by permissible stresses in the reinforcement and concrete.

C43: Prior to any concreting, the Architect must approve the bending, placing, and tying of reinforcement.

Reinforcement Placement

C44: Unless otherwise noted on the design drawings, the minimum concrete covering for reinforcement shall be as follows.

(a) 75mm concrete poured against the ground

(b) Formed surfaces exposed to the elements, water, or contact with the ground – all reinforcement, including stirrups, ties, and spirals, must have a 50mm clearance.

(c) Surfaces that aren't exposed to the elements (interior) or water, and aren't in direct contact with the ground. Stirrups, ties, and spirals must have a 12mm clearance from beams, columns, slabs, and walls.

C45: At the time of concrete placement, all reinforcement must be free of mud, oil, or other materials, such as loose rust or mill scale, that could affect or reduce the bond.

C46: To prevent displacement due to construction loads or concrete placement, all reinforcement must be supported and fastened together. Supporting concrete blocks can be used on the ground if necessary. Approved metal or plastic bar chairs and spacers must be used against formwork. Where the finished structure's concrete surface will be exposed to the elements, all accessories within 12mm of the concrete surface must be non-corrosive or corrosion-resistant. The Concrete Reinforcing Steel Institute's detailed recommendations can be found here.

Reinforcing Bars in Place (CRS1 76).

C47: Welded wire fabric shall be extended through contraction joins and construction joints, except in paving and at keyed joints in slabs on grade, and shall have lapped splices not less than the spacing of the cross wires. The wire fabric must be adequately supported during concrete placement to ensure its proper position in the slab, as described above, by laying it on a layer of fresh concrete of the proper depth before placing the slab's upper layer.

At lapped splices, vertical bars in columns must be offset by at least one bar diameter. Templates for all column dowels must be provided to ensure proper placement.

C49: The Architect will not accept any splices that are not shown on the drawings.

C50: Reinforcement shall not be bent after being partially embedded in hardened concrete unless the Architect permits it.

C51: Reinforcement bars must not be welded or spot welded for any reason.

Formwork

C52: The Contractor must notify the Architect of the general method and system of formwork he intends to use before construction begins.

C53: All formwork joints, as well as joints between formwork and previous work, must be sufficiently tight to prevent liquid from the concrete from leaking out through these joints.

C54: Formwork fixing and locating methods that result in holes through the concrete section when the formwork is removed are prohibited.

C55: No metal part of any device for keeping formwork in place must be permanently embedded within the concrete cover to the main reinforcement.

C56: The use of concrete retarders or other similar preparations on formwork surfaces requires the Architect's prior approval.

C57: Where 'extra over for fair face' is measured to concrete surfaces, the Contractor shall assume that 'sawn' formwork has been measured to those faces, and when pricing the 'fair face' item, he shall allow for the extra cost of using lined formwork or whatever other approved method he may adopt in finishing the concrete face perfectly smooth and even, free of board and joint marks and other defects. Such faces will not be allowed to be skimmed.

C58: Wrought formwork has been measured only when the exposed concrete requires a finish other than "fair face."

C59: Fixing blocks and ends of brackets, bars, and bolts, among other things, must be cast in concrete at the time of placement, and all mortices, holes, apertures, chases, and grooves, among other things, must be accurately laid out in the formwork before the concrete is poured. Without the Architect's approval, no part of the concrete works may be cut away for any such item or for any other reason.

C60: The Contractor shall obtain complete information of all sub-contractors' requirements for conduits and pipes, fixing blocks or boxes, chases and holes, and any other items to be cast or formed in concrete members from all sub-contractors, subject to the condition that failure of a sub-contractor to supply such information shall not cause the Contract to be delayed.

C61: At the start of the Contract, the Contractor shall ensure that all subcontractors are informed of the structural work programme. Before work begins, he must ensure that the subcontractor's requirements for concrete members are approved by the Architect.

C62: The Contractor shall provide written copies of the preceding two paragraphs to the subcontractor at the start of the Contract.

C63: All formwork vertical propping must be carried down far enough to provide the necessary support without causing damage, overstress, or displacement to any part of the structure.

C64: Structural props must be kept in place until the new construction is strong enough to support its own weight as well as any loads placed on it during the contract period.

C65: In two directions, each member must be supported by structural props spaced no more than 1.5 metres apart.

C66: All soffit formwork must be built in such a way that it can be removed without damaging the structural props.

C67: All beams and slabs' formwork shall be constructed with the following upward camber, unless otherwise specified on the drawings:

Getting between the supports:

At mid-span, roofing beams and slabs account for 0.2 percent of the span; floor beams and slabs account for 0.1 percent of the span.

Cantilever: 0.4 percent of the projection at the free end for beams and slabs;

C68: To prevent concrete from adhering to the forms, the internal faces of the formwork may be coated with an approved preparation, provided that the use of this preparation does not stain the finished concrete's surface. This preparation must not come into contact with the reinforcement.

C69: The interior of any section of the formwork must be completely cleared of all extraneous materials, including water, immediately before the concrete is placed in that section.

C70: The Architect shall inspect each section for formwork to structural members immediately before concrete is placed in that section.

Construction and Movement Joints

C71: The Contractor must ensure that all construction joints are arranged in such a way that the effect of concrete shrinkage is minimised. The distance between construction joints in walls and slabs should not exceed 10 metres in most cases.

C72: Before work begins, the Architect must agree on the positions of all joints not shown on the drawings.

C73: Except for horizontal joints, all construction joints must be formed with proper stop-boards, which must be fixed vertically unless otherwise directed.

74C Before the adjacent section is concreted, the laitance must be completely removed from the contract face at all construction joints. Treatment of the joint shall be terminated 12mm away from the face to be exposed in the finished works where an adjacent face of the concrete is to be exposed.

C75: Horizontal joints at exposed faces must be formed against a 12mm thick straight batten.

Reinforcement

C76: Reinforcement bending schedules will be provided, detailing the cut length and diameter or size of each bar in the works, as well as the bending dimensions and location of each bar.

C77: Before cutting the bars to length, the Contractor must make the following checks:

C78: That the reinforcement schedules for each part of the structure are provided far enough ahead of his concreting schedule.

C79: That each schedule contains the correct quantities of reinforcement as specified on the corresponding drawing.

C80: That the reinforcement grades listed in each schedule correspond to the reinforcement grades shown on the relevant drawing.

C81: All reinforcement bars must be accurately shaped to the details shown on the drawings and bending schedules in a way that does not harm the material. Bars must not be bent while hot.

C82: Prior to being placed in position in the Works, all reinforcement must be free of all loose mill scale and thoroughly cleaned to remove any loose rust, oil, grease, or other harmful matter.

C83: By using an approved method, all reinforcement must be precisely placed with the correct cover and securely fixed in the positions shown on the drawings.

C84: The bending schedules will list the mild steel reinforcement used in approved chairs.

C85: No metal part of any device used to connect bars or keep reinforcement in place against the elements, earth, or water shall remain permanently within the specified minimum concrete cover to the reinforcement.

C86: Reinforcement welding is only permitted with the architect's permission.

C87: Reinforcement shall not be bent up within the formwork unless the Architect has given his or her approval.

C88: The Contractor must ensure that any projecting reinforcement does not stain any part of the Works with rust.

Concrete pouring

C89: In approved weigh batching equipment, the quantities of cement and aggregates for each batch must be weighed separately. When cement is delivered in bags, each batch of concrete must be proportioned to use an entire bag.

C90: Concrete must be mixed in a mechanical batch type concrete mixer that has been approved. Mixing should be continued until the materials in the mixer are evenly distributed and the mass is uniform in colour. Each batch must be mixed for at least the amount of time recommended by the mixer manufacturer.

C91: The total volume of mixed materials in each batch must not exceed the mixer's rated capacity. Before the mixer drum is re-charged, each batch of concrete must be completely discharged.

C92: When mixing stops, the mixer drum must be thoroughly washed out.

C93: Concrete must be transported from the mixer to its final location as quickly as possible without segregation or loss of any of the ingredients.

C94: All concrete-transporting plant and equipment must be kept clean. When mixing stops, all containers used to transport concrete must be thoroughly washed out.

C95: Concrete transporter runs or gangways, as well as main foot traffic runs, must not be supported or allowed to bear on the fixed reinforcement.

C96: Within 30 minutes of mixing, concrete must be placed.

C97: During the placement of reinforced concrete, a competent steel fixer must be present at all times to adjust and correct any displaced reinforcement.

C98: The Contractor must keep a complete record of the Works on site, including the time and date when concrete is placed in each section of the Works. The Architect shall have access to this record at all times for inspection.

C99: Concrete must be compacted thoroughly during placement and worked carefully around all reinforcement and embedded fixtures, as well as into the formwork's sides and corners.

C100: Unless otherwise specified, all structural concrete in the grades listed on the Schedule of concrete mixes must be compacted with suitable mechanical vibrators, with internal vibrators being used whenever possible.

C101: All freshly placed structural concrete surfaces must be covered with an approved material and kept moist for 7 days, except for concrete made with rapid hardening cement, which requires only 3 days of curing.

C102: Leaving the soffit and side forms in place will be considered effective in keeping those surfaces moist.

C103: Before work begins, the Contractor must notify the Architect of the curing system and methods he intends to use for all structural concrete members.

C104: Ordinary Portland cement precast units can be steam cured at atmospheric pressure. The temperature of the units must be raised at a constant rate that does not exceed 40 degrees Fahrenheit. per hour, as well as the following requirements for curing:

C105: For measuring atmospheric shade temperature, a maximum and minimum thermometer of approved design shall be kept on site close to the works.

C106: On site, keep a daily record of the maximum and minimum morning and evening temperatures. The arithmetic mean of the maximum and minimum morning and evening temperatures recorded in each 24 hour period is used to calculate the daily average temperature.

Casting / Concreting

C107: The Contractor is responsible for ensuring that precast concrete units are stored and delivered to the job site in a timely manner. This schedule must be agreed upon in writing with the precast concrete manufacturer. Unless otherwise specified, precast units must be made of concrete Grade A.

C108: Before removing the precast units from the casting beds, the concrete must have gained sufficient strength to prevent any damage, distortion, or overstressing. The Contractor is responsible for providing all necessary lifting devices, which must be approved by the Architect prior to the units being built.

C109: The precast units must be protected from damage or surface staining during all subsequent handling, storage, and transportation. Any units that are damaged or stained may be rejected by the architect.

C110: All units must be marked in a manner and location approved by the Architect as soon as they are removed from the casting beds.

C111: Before erection, all precast units must be made available for checking of dimensions and surface finishes, and the Architect must approve them.

C112: Before beginning erection, the Contractor must submit details of his proposed arrangements for lifting and erecting precast units on site to the Architect for approval.

C113: Precast units that need to be temporarily fixed in place must be rigidly propped at a suitable point as determined by the Architect.

C114: All joint surfaces must be cleaned thoroughly. Dry-packed mortar joints are made by compacting the mortar with a steel tool in one-inch layers.

C115: Bedded mortar joints are created by placing precast units on top of a firm layer of mortar. The units must be levelled on steel shims with the top surface just below the mortar's surface level. A minimum of one inch of mortar or concrete must cover the shims. A neat cement mortar shall be spread evenly to form a thin bed just sufficient to take up any high points on the bedding surface to form thin bedded mortar joints.

C119: No treatment of any kind other than that required for curing the concrete shall be applied to the concrete faces after the formwork has been removed until they have been inspected by the Architect.

C120: All concrete faces to be exposed in the finished works shall be left as struck unless otherwise specified.

C121: All extraneous fins and similar projections must be carefully removed after inspection. To achieve a fair-face to the concrete, no render or other applied finish shall be used.

C122: During the execution of subsequent works, all concrete faces to be exposed in the finished works must be adequately protected against damage and surface staining.

Precast Units Tolerances

C123: All dimensions must be within the tolerances listed below unless otherwise specified on the drawings.

All dimensions greater than 3 metres and less than 6mm

All dimensions less than 3 metres minus 3 millimetres

Precast units must also meet each of the following tolerances, which may vary from the tolerances listed above depending on the circumstances.

Tolerances do not build up over time.

In 3 metres, a bow of 6mm is permissible.

Twist from any plane surface defined by any three exterior corners 3mm is permissible.

The finished structural elements' centre lines must be within 3mm of their correct position on plan relative to the nearest reference grid line and within 3mm of their correct level relative to the datum level, unless otherwise indicated on the drawings.

Tolerances for all sections of structural elements are as follows:

For dimensions of no more than 200mm x 3mm

For dimensions greater than 200mm but less than 3 metres) 3mm

6mm for dimensions greater than 3 metres

Surfaces exposed in the finished work must not deviate from a 1.5 metre straight edge placed anywhere on the surface by more than 3mm.

Bills of Quantities Rates

C123: All concrete work, including precast concrete, is subject to the following rates:

1) Working around reinforcement and pouring concrete into formwork.

2) Work in all sectional areas, bays, and other lengths and thicknesses (concrete work has been measured the net volume and thickness required after vibration, and the rates shall account for bulk loss due to variation).

3) Sanding all required construction joints, including formwork.

4) Leaving surfaces, beds, and backings ready for rendering of any kind.

5) Making chases, grooves, notches, mortices, holes for pipes, and other miscellaneous labours, as well as making good.

6) All tests must be completed.

7) Security.

The following items must be included in the bar reinforcement rates:

1) Any length of bar.

2) Extra labour is required to form links, stirrups, binders, and special spacers.

Bends, hooks, tying wire, distance blocks, and ordinary spacers are all included in this category.

4) Security.

The following items must be included in the fabric reinforcement rates:

1) Any type of cutting, notching, bending, tying wire, and distance blocks are all acceptable.

2) The extra material at the laps (Unless otherwise specified, the end laps for square meshes shall be 300mm with 150mm side laps).

The following items must be included in all formwork rates:

1) Any description of cutting

2) Notching, drilling, boring, perforating, and other similar tasks.

3) Spacers, distance pieces, sleeves, and other similar items

4) Formwork to soffits of slabs of any thickness and strutting to any height.

5) Unless otherwise stated, the formwork and supports are only for temporary use and waste.

6) Security.

The following items must be included in the precast concrete rates:

1) All moulds and casings.

2) Hoisting, bedding, or positioning in place.

3) If necessary, pointing around and joining units.

4) Applying a fair finish to all exposed surfaces

5) Extra reinforcement for the sole purpose of reducing the risk of breakage during handling.

6) Safety and security

SECTION D BLOCK WORK, BRICKWORK AND MASONRY

D1: For blockwork, brickwork, and stone masonry, use the cement specified in Section C – Concrete Work.

D2: Natural sand, clean sharp coarse grit pit, or river sand shall be used in the mortar preparation.

D3: Lime shall be freshly burned stone-lime delivered in large lumps and slaked on site; hydrated lime shall be used in the absence of slaked lime.

D4: Water must meet the requirements of **Section C – Concrete Work.**

D3: If the Contractor wishes to manufacture blocks off-site or use blocks from a third-party supplier, he must notify the Architect and provide him with information about the manufacturer or supplier. No block supplier will be approved unless the Contractor obtains a written guarantee from the supplier that blocks will be manufactured in accordance with the contract's requirements, as well as the supplier's agreement to the Architect inspecting his factory during the block making process.

D4: Samples of all sizes of blocks required in the Works must be submitted to the Architect for approval before any work begins.

D5: Hollow cored blocks with a minimum shell thickness of 45mm are required.

D6: Blocks are made up of one part cement to six parts sand, mixed with just enough water to bind the mixture under pressure. Within half an hour of adding water to the mix, the mixture must be well rammed and consolidated in the mould and made into blocks.

D7: The nominal face size for rendering purposes is 440 x 215mm keyed.

D8: Blocks must be cast true to shape, consistent in size, smooth-faced, free of flaws or blow holes, and with clean, sharp arises.

Blocks' Curing

D9: After being removed from the machine, the blocks must be placed on pallets in separate rows, one block high with a space between them, to mature in the shade for at least 24 hours and sprayed with water.

They can then be removed from the pallets but not stacked or taken out of the shade for another four days, during which time they must be sprayed with water at regular intervals. After that, the blocks should be stacked in the shade, not more than five blocks high, and allowed to dry. No blocks may be placed in any part of the structure until they have matured for at least 21 days in this manner.

D10: Blocks must be able to withstand a pressure of 2.8N/mm2 (400lbs/sq.in) 28 days after manufacturing, and the Contractor must allow samples and tests to determine compressive strength as directed. If any sample fails to reach the required strength, the Architect/S.O. may reject the batch containing the sample and order the Contractor to break up or remove the entire batch from the job site.

Mortar

D11: Mortar must be made up of cement and sand (1:6), as well as an approved plasticizer, and must have a strength equal to that of the blocks.

D12: It must be used within one hour of mixing and not mixed with any other mortar after it has started to set, nor with any of the previous day's mixing.

D13: Appropriate stages must be set up to receive the mortar when it is ready.

D14: All blockwork must be plumb, true to line, and level throughout. All vertical and bed joints must be mortared solidly. Blockwork must be solidly built around joists, lintels, bolts, pipe sleeves, and other structural elements. The exposed edges of the blocks must be square and even when cut.

D15: Stretcher bond blockwork is required. The walls must be built in a consistent manner, with level courses and perpends maintained. One course must be 230mm in height, with joints measuring approximately 10mm. At any given time, no portion of the walls may be built one metre higher than another.

D16: In dry weather, all blocks must be thoroughly soaked before use, and the tops must be thoroughly wetted before work can resume.

D17: During the course of the work, the entire width of the tops of the walls that are not being worked on must be protected. During the construction process, the tops of walls must be covered.

D18: As the work progresses, hollow blocks filled with solid should be filled in each course. Filling concrete should be Grade D, as described in Concrete Work.

D19: As the work progresses, fair face surfaces should be joined in mortar and then thoroughly cleaned with a wire brush. Cleaning should begin at the top of the walls and work its way down, with cracks, holes, and other defects being carefully identified and filled and pointed with mortar.

D20: The joints on surfaces that will be used for rendering, etc., must be raked out to a depth of 9mm.

D21: Movement jointing must follow the guidelines outlined in Concrete Work.

D22: Where BLOCK WORK butts against vertical concrete faces, dovetailed steel strips of an approved make shall be fixed in the formwork to receive bonding ties or butterfly ties made of 6mm steel rod, or approved brick strip reinforcement shall be cast in the concrete at not more than 0.3m centres and to suit courses for building into the horizontal joints of BLOCK WORK.

Chasing

Horizontal chases are not allowed in walls. Vertical chases must be solidly bedded in cement mortar where they are permitted.

Bill-of-Materials Rates

D23: The following items, as well as making good, will be included in the rates for work in this section:

(1) If not otherwise specified, rake out joints on both sides to serve as a key for plaster.

(2) Any sort of cutting

(3) Making good grooves, chases, mortices, and sinking, as well as cutting or forming grooves, chases, mortices, and sinking.

(4) Cutting and pinning the ends of all lintels, steps, timbers, tubular rails, brackets, steel sections, and other similar materials.

(5) Providing a brickwork sample panel as requested.

(6) Inspection and defence

SECTION E: ROOFING

E1: The entire roofing system must be secure and watertight. The Contractor must ensure that the roof is adequately protected from the following trades' movement and that such movement is kept to a minimum.

E2: All corrugated sheets and fittings must be installed in accordance with the manufacturer's instructions and in such a way that the roof structure can move slightly without causing damage to the roof covering. Unless otherwise stated, the sheets will be fixed with the recommended side and end laps between sheets, and the rates will include them.

E3: Sheet sizes must be such that all horizontal laps occur over a purlin immediately. All fixing holes must be drilled through the corrugations' tops.

E4: Purlins must adhere to the roof's profile and maintain an even slope across all falls. Before sheeting begins, the Contractor shall provide a straight gauge and the purlins shall be checked in the presence of the Architect/S.O. and the roofing sub-contractor, if such sub-contractor is required in the contract.

E5: Before sheeting begins, any defective purlins must be removed or otherwise corrected as directed by the Architect/S.O.

Roofing Sheeting with Corrugations

E6: Unless otherwise specified, the sheets must be correctly mitered and laid with at least 230mm head laps and one and a half corrugation side laps. The laps' exposed edges must face away from the prevailing winds.

E7: From eaves to ridge, all courses must be corrugated in straight lines; the mitred corners of sheets must be properly weathered by the overlapping corrugations of the sheets above.

All fixing holes in the crown of the corrugations must be drilled rather than punched, and all cutting must be done neatly with a hand saw.

E8: Hooks, bolts, seam bolts, and drive screws with slotted heads shall be used to secure the sheeting and fittings where and as required. For asbestos cement, the bolts and screws shall be galvanised mild steel, and for aluminium sheeting, the bolts and screws shall be of approved pattern and dimensions and shapes to suit the purlins, etc., to which the sheeting shall be fixed.

Bolts and screws must have a minimum diameter of 8mm and a minimum diameter of 6mm. All bolts and screws must have galvanised steel diamond washers, aluminium washers, or lead cup-shaped washers bedded on felt-asbestos composition washers or plastic washers, and no fixings must be closer than 38mm to a sheet's edge.

E9: Where plastic seals are required, unless otherwise specified, the laps in the roof covering shall be sealed with approved bituminous mastic as recommended by the manufacturer.

Corrugated Aluminium Sheeting

E10: Corrugated aluminium sheeting from Flag Aluminum, Alumaco, or another approved manufacturer must be obtained. The sheeting must be 0.7mm (22SWG) and have a stucco mill finish unless otherwise specified.

Capping and flashing made of aluminium

E22: Aluminum flashings and capping must meet BS.1470 Super Purity grade requirements. With a milled stucco finish. They must be 0.9mm thick unless otherwise specified.

E23: Ends of flashings must be lapped a minimum of 150mm.

E24: Before installing flashings and capping, the underside must be coated with bitumen.

Bills of Quantities Rates

E25: The following items are included in the rates for flashings and the like:

(1) Laps and seams in the middle

(2) Passings at intersections, angles, and ends

(3) Before fixing, coat the underside in bitumen.

E26. Capping rates must also include rates for ends, angles, and intersections.

Protection

E27. All roof coverings and flashings shall be priced to include protection.

SECTION F CARPENTRY AND JOINERY

Quality Timber

F1: All timbers must be of approved quality, well seasoned, dry, and defect-free, and must comply with BS.4978 and BS.881 nomenclature.

F2: Mahogany, Opepe, Black Afara, Agba, or other approved timber shall be used for carpentry work and must comply with BS.4978.

F3: For joinery work, the wood must be Mahogany, African Walnut, Iroko, Agba, or other approved wood that meets the requirements of BS. 1186, Part 1, and the workmanship must meet the requirements of BS.1186, Part 2, sections 1 and 2.

The wood must be well-seasoned hardwood of the highest possible quality, free of all defects, selected for grain and colour, and approved by the Architect/S.O as suitable for the intended use.

Timber preservation

F4: All timbers must be treated with an organic solvent or nonstaining solution in accordance with BS.1282. The latter should be used on wood that will come into contact with render or other decorative surfaces and will need to be painted. All timber to be embedded in blockwork or concrete, as well as the backs of window frames, must be treated. After the timber has been cut to size and shape, the preservative should be applied. If preservative treatment is performed off-site, certification must be presented to the Architect; if treatment is performed on-site, prior notification to the Architect/S.O. is required.

Wrought Face

F5: Unless otherwise specified, all exposed faces of timber must be wrought. Wrought-timber sizes listed in Bills of Quantities are finished sizes.

Preparation

F6. Timber preparation should begin at the same time as the rest of the work, and should continue until all of the woodwork is prepared and stacked in such a way that air can freely circulate between them. After receiving the detailed drawings, all framed work must be cut out and framed together as soon as possible, but not glued or wedged up until ready for fixing.

F7: As soon as the Contractor has possession of the site and detailed drawings, the joinery work will be cut out, skeleton framed, and stacked on site. It must be carefully stored and protected from the elements, but not wedged up until it is needed for installation in the building.

Before being wedged up, any parts that wrap or develop shakes or other defects must be replaced with new ones.

Fixing

F8: The provision of all necessary glue, nails, screws, and other fixings to adequately secure the timber in an approved method and as directed shall be included in the fixing and framing of timber generally described hereafter.

Framing

F9: Carpentry work must be properly framed up and finished in a professional manner. Plates must be halved at corners, and joints in the length of purlins, rafters, and the like must be scarfed and spiked for at least three times the depth of the timber and wrapped in hoop iron.

Scarfing

Wherever possible, joints should be made over a point of bearing. At all angles, wall plates must be bedded perfectly level with halved joints and dovetailed lapped joints. They'll be fastened to the walls with 13mm ragbolts spaced at 1.2m intervals.

Fabrication

F10: All joinery work must be properly laid out, framed, and executed in accordance with detailed drawings, and finished in a workmanlike manner.

F11: All work must be crafted and finished with a clean, smooth, and even surface, with all exposed external angles to frames, etc. rounded.

F12: Allow for keeping the entire joinery project, including door leaves, clean for staining or polishing until the Architect's decorative treatment requirements are known.

F13: All glued joints must be properly feather-tongued, and all mouldings to framing must be mitered at angles as needed.

F14: Adhesives must meet BS requirements. For both internal and external use, use grade W.B.P for plywood and BS.1204 for other wood.

F15: All plugged members must be secured with hardwood or other approved plugs or compound. The plugs shall be spaced at such intervals as to provide adequate fixing, subject to the Architect's approval.

Flush Doors

F16: Flush doors must be built in accordance with the material requirements of BS.459, Part 2 and must be covered on both sides and have a hardwood lip all around. Skeleton framed doors must have enough solid blocks for ironmongery to be installed.

F17: The rates for attaching doors to metal frames must include the cost of attaching lift-off half-butts that come with the frames.

Plywood

F18: Plywood must be of the highest quality, weather-resistant, boiled and bonded, mahogany-faced on both sides, and purchased from an approved manufacturer.

F19: The carpentry and joinery rates must include the following items:

(1) Any lengths of wood

(2) All types of labour

(3) Any and all types of joints

(4) Raising and lowering trusses to any height

(5) Using an anti-termite preservative that has been approved.

(6) Before fixing to concrete or blockwork, prime the back of the wood.

Bills of Quantities Rates

F20: The carpentry and joinery rates must include the following items:

(1) Any and all labours and joints of any kind.

(2) Seasoning and treating with the approved preservative.

(3) Applying a coat of primer to the back of the wood before attaching it to walls or other surfaces.

(4) Security

Wrought Iron

F21: All ironmongery, as well as door and cupboard furniture, must be ordered from the specified supplier. The contractor must provide ironmongery samples in advance.

F22: Screws of the same metal and finish as the fittings must be used to secure ironmongery. All screws that have been damaged while driving must be withdrawn and replaced with new ones.

F23. Oil and adjust all locks and other fittings, and leave them in good working order.

F24. Prior to handing over the works, or any section of the works, the Contractor shall attach wooden tabs to each set of keys, clearly identifying which lock they belong to, and deliver all sets of keys to the Architect mounted on a suitable plywood keyboard.

F25: All locks must come with two keys unless otherwise specified. Except for a master key if specified, no two locks shall be able to be operated by the same key.

F26: The rates for ironmongery fixing shall include the supply of matching screws, the temporary removal of all ironmongery prior to painting and subsequent re-fixing, and protection.

SECTION G STRUCTURAL STEELWORK

Quality of steel

G1: For structural purposes, all steel plates, sections, bars, and hollow sections must comply with BS.4360, Part 2 grade 43A.

G2: Upon request from the Architect/S.O., the Contractor shall provide certificates from the manufacturer certifying that the chemical constituents of the steel meet the British Standard.

G3: For the appropriate stress requirements, the steel grade shall be that listed in Table 2 of BS. 449.

G4: All structural steel sections must meet the requirements of BS.4, Parts 1 and 2, as well as BS. 4848, Part 4.

G5: All steel must pass Lloyds inspection and be free of mill scale rust and pitting.

G6: The weight of mild steel shall be as specified in BS.648 for measurement purposes.

Bolts and Washers

G7: All bolts and nuts must be black steel bolts and nuts that meet the dimensions of BS.4190 and the screw threads of BS.84. Lewis bolts, hook bolts, and other fasteners must comply with BS.1494. Unless otherwise specified, all bolts and nuts must have hexagon heads and Whitworth standard threads. BS.4320 specifies the requirements for washers.

Tolerances

G8: In general, the accuracy of dimensions and position of attachment and fabrication shall be + 3mm, or whatever greater accuracy the Contractor deems necessary to meet the erection tolerances listed below.

G9: Individual members must be erected to within + 3mm of the dimensions and levels shown on the drawings in terms of accuracy, position, plumb, level, and square. In 30m, the overall length, width, and height tolerances must be +6mm.

Fabrication Generally

G10: Fabrication shall generally be in accordance with BS.449, unless the following clauses amend or add to it.

G11: To achieve the tolerances stated, plates, bars, joists, and sections must be true to form and accurately straightened, planed, or shaped as needed.

G12: Steelwork must be fabricated by an approved specialist firm, with as much fabrication as possible taking place in the manufacturer's works.

G13: Where members' ends bear compression, they must be cold sawn and then machined to ensure that the loads are evenly distributed throughout the section.

G14: Cold or hot sawing can be used when the ends of members are not in compression or when the member needs to be cut for notices or holes. Machine gas cutting is only permitted if the Architect/S.O. has given his or her written approval. Under no circumstances will manual gas cutting be permitted.

G15: Burrs left by the cold friction or hot saw must be removed, and gas cuts (where gas cutting is permitted) must be dressed to a neat and workmanlike finish, free of oxidised metal.

G16: Fitted stiffening angles or plates to brackets or section flanges must be accurately shaped to fit the profile of the stiffened member.

Connection with Bolts

G17: All steel holes must be drilled precisely to templates. Before the parts are assembled, burrs and arrises must be removed from the edges of holes, and holes must not be punched without the permission of the Architect/S.O. All countersinking must be concentric and at the proper depths.

G18: Bolts in bolted connections must be of such length that when tightened, a minimum of 6mm of each bolt shank protrudes through the nut.

G19: Drift pins will be allowed only for connecting various parts of the structure, etc., and will not be allowed to distort the work or enlarge the bolt holes.

G20: All bolts that pass through wood must have washers under the heads or nuts that bear on the wood.

G21: Welding shall be done in accordance with BS.449 and BS.2642.

G22: For arc welding, mild steel electrodes must meet BS.639.

Welding on the Job

G23: Site welding is prohibited unless the Architect/S.O. has given written permission. The Contractor shall propose methods of on-site welding inspection to satisfy the Architect/S.O of the Work's integrity.

Drawings

G25: From the Architect's drawings, the Contractor shall prepare erection scheme drawings, calculations, and shop details for approval.

Painting before Fabrication

G26: All fabricated members must be shot blasted in accordance with Swedish Standard S/S 055900 1962: To remove all mill scale, rust, foreign matter, and achieve a white metal condition over the entire surface area, use Sa 21/2 low profile; or any other method as directed by the Architect/S.O. Oil, grease, and dirt must be removed from all loose material and dust, and the area must be completely dry.

G27: One coat of M.E.P Metal Conditioner (see Painting Preambles) must be applied within 4 hours of shot blasting, strictly following the manufacturer's instructions and to the thickness recommended.

G28: Welded or bolted members' contact surfaces must be prepared and painted with one coat of metal conditioner before assembly.

After Fabrication, Painting

G29: Any damage to the metal conditioner, including that caused by weld splatter, must be repaired after welding. Steel members must have an additional coat of metal conditioner applied by roller or brush to all edges, arises, and exposed ends.

G30: A 24-hour period must elapse before the Steelwork is transported to the job site to allow the metal conditioner to cure.

Surfaces that have been painted need to be protected.

G31: The Contractor is solely responsible for the paint system's subsequent protection at the job site, during transportation, storage, and erection.

G32: Protective measures must be taken during handling operations to avoid chains or other handling appliances coming into contact with the paint system. Suitable wooden spacers must be used to keep each item separated from its neighbours while in storage and during transportation.

After Erection, Painting

G33: After erection, the Contractor must repair any damage to the metal conditioner by cleaning all damaged areas back to clean metal, ensuring a good bond to the existing primer, and painting an additional coat of metal conditioner.

G34: If the Contractor wants to use a different Paintwork Specification, he must submit details to the Architect/S.O for approval in writing. Any alternative Specification must provide at least the same level of protection as the one described above.

Testing and Inspection

G35: All on-site tests and inspections must be conducted in the presence of, or as directed by, the Architect/S.O.

G36: All tests specifically required in this contract shall be included in the Contractor's rates.

G37: In the event that the Contractor fails to comply with the contract, he will not be paid for any special tests that the Architect/S.O may require.

G38: In addition to clause 55 of chapter 5 of BS.449, the Contractor shall, upon written request from the Architect/S.O, arrange for the testing of any permanent works materials and/or workmanship to be carried out either at his works or on site.

G39: The Contractor will be paid at agreed-upon rates for any other special tests requested by the Architect/S.O, unless the results of the tests show that the material or workmanship is defective in some way, in which case the Contractor will be responsible for the entire cost of the testing. If the testing reveals that the material or workmanship is satisfactory, the Architect/S.O. may include the cost of the testing.

G40: The Contractor is responsible for the entire cost of tests performed under Chapter 6 of BS.449, as well as tests and test certificates used to verify the chemical composition of the steel.

G41: It is emphasised that the Contractor is solely responsible for the inspection and quality control of all materials and workmanship to be incorporated in the permanent works at all times, and that he must not respond in any way to any inspection or testing conducted by the Architect/S.O.

Site-based work

G42: The Contractor is responsible for all unloading, handling, erection, and painting work, as well as providing all necessary plant and equipment, including any scaffolding.

G43: After unloading, all materials must be kept in a designated area approved by the Architect/S.O until they are needed for erection.

G44: The Contractor is responsible for the stability of the steel trusses during erection and must take the necessary precautions to maintain the steelwork's correct position in the works by properly strutting and staying it during erection and when it is fixed. All temporary strutting, etc., shall be removed after erection and the setting of any grout to beam seatings, etc., so that the steelwork's stability is achieved according to the design method.

G45: The erection of the steelwork may be carried out in sections and thus offered for inspection with the prior written approval of the Architect/S.O.

G46: At least 21 days before any erection, the Architect/S.O must receive written details of the method and sequence of erection.

G47: When the work's layout is presented to the Architect/S.O. for inspection, it must be completely finished and checked by the Contractor. The Contractor's request for inspection must be communicated to the Architect/S.O. 24 hours ahead of time.

Only after the Architect/S.O has approved the setting out of the trusses may the grouting up of beam seatings, etc. begin. The Architect's/approval S.O.'s has no bearing on the Contractor's contractual obligations in this area.

Protection

G48: Until the Architect/S.O. approves the structural steelwork and all painting thereto, the Contractor is responsible for protecting it and all painting thereto.

G49: After erection, the Contractor shall leave the steelwork clean and shall perform any necessary cleaning operations at his own expense.

In general,

G50: Each steel section size is listed separately, along with the number of pieces included in the weight.

G51: All steelwork was measured according to the rules for framed steelwork, with the exception that all fittings and connections to steelwork were measured separately and are not included in the weight of the item to which they are attached. Splice plates, bolts (either shop fitted or site fitted), rivets, gusset plates, base plates and caps, and other fittings in connection shall be interpreted in the broadest sense.

G52: Lattice girders have been grouped under a specific heading, with each member listed separately.

G53: Where applicable, high-strength friction grip bolts are included in the Fittings' weight and are not listed separately.

G54: The structural steelwork rates must include the following:

1) Hoisting to the required heights. Steel wedges will be used to wedge up and level the roof trusses.

2) Priming prior to delivery to the job site.

3) All labour in drilling holes for all trades, bends, ramps, and other similar items.

4) Welding and bolting on-site at all joints and intersections as specified.

5) Protection.

SECTION H METALWORK

General

H1: The material and workmanship standards shall not be less than the current British Standards, as referred to in these Specifications, or any similar National or International Standard approved by the Architect.

Steel Sections

H2: Hot rolled sections must comply with B.S. 4 (Structural Steel Sections, Part 1) and B.S. 4848 (Hot Rolled Structural Steel Sections, Part 4)

H3: Cold formed sections must meet the requirements of B.S. 2994 'Cold Rolled Steel Sections.'

Aluminum is a metal that is used to

H4: Aluminum shall be manufactured in accordance with the following British Standards in the category of "wrought aluminium and aluminium alloys."

'Plate Sheet and Strip' (BS 1470)

'Drawn Tuibe,' B.S 1471

'Rivet, Bolt, and Screw Stock,' B.S 1473

'Bars, Extruded Round Tube and Sections,' B.S 1474.

H5: The aluminium sections must conform to B.S. 1161 'Aluminium and Aluminium Alloy Sections,' with the alloy most appropriate for the member's location.

H6. Where aluminium comes into direct contact with concrete or concrete blocks, it must be protected by first painting with a self-etching primer, or the concrete must be painted with two coats of bitumen to isolate surfaces in contact, or it must be spaced apart by plastic packing.

Iron and Steel Galvanization

H7: Unless otherwise specified, galvanised steel and ironwork of any kind must be pickled in dilute hydrochloric and acid, then washed, fluxed, and stored before being zinc-coated by dipping in molten zinc baths.

H8: All articles must be immersed in the bath for only as long as it takes them to reach the bath's temperature, and they must be withdrawn quickly enough to achieve a coating of 500 grammes per square metre of surface. In all cases, galvanising should be done after the chipping, trimming, filling, fitting, or bending has been completed.

Aluminium with anodized finish

H10: Where shown, the aluminium alloy members will be anodized in accordance with B.S. 3987, "Anodized Wrought Aluminum for External Engineering Applications."

H1: All aluminium sections, panels, accessories, and components must be anodized with a 25 micron thick coating of 'Calinal.'

H12: All anodized aluminium sections, components, rails, bars, mullions, and the like shall be protected by an approved transparent film sprayed on the completed unit before despatch from the manufacturer and removed as and when instructed by the Architect/SO upon completion of the works. The transparent film must be removed using a solvent or another approved method that does not harm the frame finish or adjacent work.

Metalwork Units Manufacturing

H13: All metalwork components must be properly fabricated, cut, wedged, and assembled in accordance with industry standards. All contact surfaces must be square and true and fit snugly. Welded connections must be made in the shop, and no site welding is permitted unless the Architect/S.O. has given specific approval in writing. Machine welding is preferred, and all welding must be performed by competent and qualified welders who follow all applicable regulations.

These Preambles contain British Standards and clauses.

Unit Transportation

H16: All metalwork units must be properly prepared, bundled, and packed for shipment and transportation, with appropriate precautions taken to ensure that the paintings and protective treatments are not damaged during shipment and transport to the Site.

Mechanical Services Insert Fixing

H17: The concrete inserts needed to support metal components, mechanical and electrical services equipment, pipes, and other items must be of the type specified. Those inserts must be continuous or short in length as needed, and they must include self-locating nuts and caps. While being cast into the concrete structure, the inserts will be filled with a polystyrene filler. Each insert must be the appropriate size to support the imposed load.

Panels for Windows and Walls

H18: Windows and wall panels must meet the stated specifications in general. The entire system for the external elevations must be approved and meet the following requirements:-

1. Vertical mullions with 90mm deep sections weighing 2.80 kg per linear metre

2. 1.60 kg per linear metre horizontal frame with 45mm deep section.

3. Aluminum U-channel top sub-frame to absorb building tolerances.

4. The bottom sub-frame must be made of galvanised steel or aluminium.

H19: The external elevation window and panel system must be able to withstand a minimum wind pressure of 120 kg/sqm.

H20: The window system for the internal elevations must be approved and meet the following requirements:-

H21: All aluminium sections, panels, and accessories must be anodized the same colour, which will be determined later.

H22: Stainless steel shall be used for all other accessories and fixing materials.

H23: The Architect/SO must approve all working drawings for all glazing and panel details.

H24: For approval purposes, the Contractor shall provide samples of all materials intended for the external and internal elevations window element. Within 30 days of receiving a letter of intent, such samples must be provided.

H25: The Contractor is to provide and erect a full mock-up of standard external and internal elevations on the existing concrete mock-up structure in the positions indicated by the Architect/S.O.

For the purpose of approval, window elements are used. The mock-up must be built within 90 days of the samples being approved.

In general,

H26: All metalwork and wrought iron work must be completed according to the drawings' details and dimensions.

H27: Fixing holes in metal must be drilled clean and burrs removed. All countersinking must be concentric and done to the proper depths.

H28: Bolts must have washers under the nuts and tapered washers under the heads and nuts if they bear on a tapered surface. Each bolt's threaded portion must protrude at least two threads through the nut.

H29: All iron, both cast and wrought, must be of approved origin, free of all defects, castings, and other impurities, and be perfectly true to shape, clean, and sharp.

H30: Welding must be done in accordance with British Standard 1856. Unless otherwise specified, welded joints, angles, and other components shall be considered for grinding or filling to a smooth finish.

H31: Skilled operators must weld and braze using either gas or electrically operated equipment, depending on the nature of the work.

H32: Unless otherwise specified, welds must be continuous fill at welds.

H33: Welding rates must include all labour, materials, appliances, and on-site and off-site welding.

Bills of Quantities Rates

H34: Metalwork rates must include the following items:

1. Hoisting to the required heights

2. All drilling labour for all trades, bends, ramps, and so on.

3. Welding and bolting on-site at all joints and intersections as specified.

4. Safety and security.

Completion

H35: Protect metals from cement and dirt, clean all window louvres, etc., lightly grease all moving parts, and make sure everything is in good working order.

SECTION I FINISHES – FLOOR, WALL AND CEILING FINISHES

General

J1: The quality of materials and workmanship must meet or exceed the current recommendations:

'The flooring and slab flooring,' B.S.C.P 202

Part 2 of B.S.C.P. 204, 'In-situ floor finishes,'

B.S.C.P 211 'Internal Plastering,' B.S 5492, B.SC.P, 212, Part 2 B.S. Part 1, for wall tiling, B.S.C.P 5262 'Code of Practice for External Rendering finishes, and the relevant British Standards as referred to in these specifications.

J2: Any inconsistency between the above recommendations and the following specifications must be brought to the Architect's/attention. S.O.'s

Materials

J3: Ordinary Portland cement shall be used, as previously described in 'Concrete Work.'

J4: White cement shall be Portland White Cement obtained from a manufacturer or their accredited agents approved by the Architect/S.O, and shall comply with B.S 12 Portland Cement to the extent applicable (Ordinary and Rapid Hardening).

J5: Colored cement must be of a specific type and colour, and pigments must be evenly distributed throughout. Pigments will not account for more than 5% of the cement's weight. It must be used in accordance with the printed instructions provided by the manufacturer.

J6: Sand shall conform to ES C.D5 003 or B.S 1199 and shall be as specified under 'Concretework.' The Architect/SO must approve the sand and it must be free of impurities.

J7: Tiles must come from approved sources and meet the following requirements:-

1. Ceramic floor and wall tiles that meet BS 1286 and 1281 standards.

2. Tiles made of cement 302 ES C.D3

3. Floor tiles and skirting in terrazzo D3, 303, ES

4. Follow the manufacturer's instructions for marble and granite tiles.

a few examples

J8: All finishing material samples must be provided prior to ordering, and the colour pattern and quality of all tiles must be approved by the Architect/SO.

body of water

J9: Water must be as specified in the 'Concrete Work' section.

FINISHES APPLIED ON THE SPOT

Cement and Mortar (Composite Mortar)

J10: The components or mortar must be accurately measured by volume and thoroughly mixed for at least two minutes until they have a uniform consistency and colour.

In the case of compo mortar, lime and sand must be mixed together before adding cement, and only a small amount of water must be used to achieve a workable consistency.

J11: Mortar batches must be adequate for use before partial setting occurs; re-tempered mortar must not be used.

J12: Cement mortar components for rendering and flooring Unless otherwise specified, screeding shall be cement and mix proportioned 1:3 by volume.

Rendering Surface Preparation

J13: After the chases for services have been cut, the services installed, and the chases have been made good, the rendering will be applied.

J14: Before beginning the rendering process, the surface to be rendered must be thoroughly cleaned with clean water.

J15: The Architect/S.O. may direct the installation of wire mesh for reinforcing the rendering where joints exist between two surfaces of different material characteristics.

J16: Before beginning the first coat of rendering, both external and internal rules must be run, and all rendering must be plumb.

J17: Before applying the first coat of render, dubbing out the uneven surfaces of the walling is required.

Preparing the Surface for Pointing

J18: Rake the joints of masonry surfaces to receive pointing to a depth specified by the Architect/SO, and remove all mortar.

J19: Before applying pointing, the joints must be thoroughly wetted.

J20: Any joint irregularities must be chased and made good. Screeding Surface Preparation

J21: Clean concrete floors that will receive cement screed thoroughly with clean water.

J22: The concrete surfaces must be kept damp for at least 48 hours.

J23: The prepared surface must be kept clean and free of trash, as well as the trades that follow.

Rendering's Application

J24: Using a trowel and a wood float, apply mortar to rendered surfaces according to accepted procedures.

J25: Use a wood or steel float to apply a single coat of rendering that is fair and smooth.

J26: After the final coat is applied, the cement rendering must be wetted for seven days.

J27: In wall rendering, all rises and angles must be straight and true, with all rises to reveals or openings slightly rounded.

J28: Waterproof rendering must include an approved liquid waterproofing agent and be applied according to the manufacturer's written instructions.

Pointing Techniques

J29: Mortar should be applied to pointing joints with a trowel and guide steel that is smaller than the joint size.

J30: Enough mortar shall be packed into the joints to create a recess of the specified depth.

J31: The wall must be kept wetted for at least three days after the cement mortar has set.

J32: Acid wash the finished masonry before cleaning it with water.

Cement Screed Application

J33: Any standing water or foreign matter on the concrete surface must be removed.

J34: Following the application of the slurry bond coat, a slurry bond coat must be applied immediately.

J35: After the slurry bond coat has been applied, the mortar bed should be applied immediately.

J36: Timber beads for running of rules shall be placed at 2m intervals to ensure proper level or falls as the case may be.

J37: The specified mortar must then be well compacted between the rules, levelled off with a straight edge, and worked on with a steel trowel.

J38: The cement screed must be smooth, even, and true level, or fall to the required levels.

J39: The finish of the cement screed must be adequately protected from subsequent trades.

J40: Keep the cement screed finish wet for at least seven days to prevent it from drying out too quickly.

Tyrolean Rendering

J41: Three coats of Tyrolean rendering are required. The first coat should be made up of one part cement to four parts sand, applied 15mm thick and smoothed out with a wood float.

J42: One part cement (mixed in a ratio of three parts ordinary cement to one part coloured cement) to two parts sand by volume, mixed dry in a ratio of three parts mixture to one part water by volume. A Tyrolean applicator should be used to apply the finishing coat to a total thickness of 7mm. Straight and true rises and angles are required.

Granolithic Paving

J43: Granolithic paving must be at least 50mm thick and be made up of one part cement and 2.5 parts granite chippings by volume. The chippings must be 6.4mm in diameter, dust-free, and graded so that 20% of the fine material passes through a 4.76mm sieve.

J44: The materials must be thoroughly mixed for at least 1.5 minutes. Before each batch is introduced, it must be completely discharged from the mixer. After the mix has left the mixer, no additional water or other material may be added.

J45: The granolithic mix is laid over to the required thickness and levels, and fully compacted with a hand tamper weighing not less than 7.5 kg per metre run or a power float, as directed by the manufacturer.

J46: Where the paving must remain monolithic with the slab, the mix must be applied to the slab while it is still green (i.e. within 4–5 hours of laying the slab).

J47: It must be laid in bays no larger than 1 square metre, with ebonite dividing strips.

J48: Complete compaction is required along the edges and in the corners. After the granolithic has been laid out, levelled, and compacted completely. It must be trowelled at least three times at intervals of 6–10 hours to produce a uniform, hard, and abrasion-resistant surface. Before trowelling, any laitance should be removed.

J49: After the paving has matured, it should be polished with a carborundun polishing machine to a fine, smooth, true surface that is free of uneveness and blemishes.

J50: A sample room will be built first, and once approved, it will be used as the standard for the rest of the work. Terrazzo Paving In-Situ

J51: The overall thickness of in-situ terrazzo paving shall be 50mm, with an in-situ topping of not less than 25mm laid on a screeded bed consisting of one part cement to three parts sand by volume, while the screeded bed is still plastic.

J52: One part coloured cement to two parts marble chips shall be used for the topping. The chips must be graded from 22mm to 3mm in size and must be free of dust or fine material.

Technical Specifications

J53: Care must be taken when mixing the materials to ensure uniformity. The aggregate must be thoroughly mixed before adding the coloured cement and blending everything together while it is still wet. It is important to avoid heaping the mixture because this will separate the larger chippings. As the mixing continues, add the water in a fine spray. It's critical to keep the water-to-cement ratio as low as possible to avoid excessive shrinkage during drying and reduce the risk of cracking, and it should be as low as possible for proper workability.

J54: It must be laid and finished in bays no larger than 1m x 1m or as directed, with the use of ebonite strips. Tower to a dense, even surface free of holes and blemishes, being careful not to bring too much cement to the surface.

J55: It must be ground smooth with an approved abrasive stone, with all holes filled with matching cement, and then machine polished with lead pads or other approved methods to achieve a highly polished finish.

J56: Hair-cracks or grazing visible with a lens under bright light, visible shrinkage cracks at the dividing strips and corners of sections, surface powdering, or any pitting due to aggregate coming off the surface are prohibited.

J57: A sample room will be built first, and once approved, it will be used as the standard for the rest of the work.

J58: The rates will include free colour selection of cement and marble chips, as well as the provision of a sample room.

Cement and Sand Paving

J59: Cement and sand paving shall be prepared and laid in the same manner as granolithic paving, with one part cement to three parts sand.

Screeds (Beds and Backing)

J60: One part cement to three parts sand shall be used in the beds under floor finishes and the backing behind wall finishes. Only enough water should be added to make a stiff mix after the materials have been thoroughly mixed dry.

J61: Before applying the bed or backing, the surface of the concrete floor slab or wall should be rough and well wetted.

J62: Beds should be laid on the concrete floor slab to a thickness of at least 25mm or as needed to achieve proper finished levels. At the edge and corners, full compaction must be maintained.

J63: Finish the beds with a steel trowel, wood float, or hair brush to create a light and even texture that will allow the final surface finish to adhere properly. Depending on the final surface finish to be applied, beds are described as screeded, trowelled, or floated.

J64: Unless otherwise specified, backings must be applied in one coat as described for rendering to a finished thickness of 12mm.

J65: Keep the beds and backing moist and out of direct sunlight until the hardening and curing processes are complete.

Concrete Sub-base

J66: Prior to the application of paving or beds, the concrete sub-floor must be prepared by mechanically or manually removing laitance to expose clean coarse aggregate particles. All dust from hacking and other sources must be removed, and the sub-floor must be thoroughly wetted with clean water.

Non-slip Surfaces

J67: Where a non-slip surface is required on paving or beds, 1.25 kg carborundum powder should be sprinkled on at a rate of 1.25 kg per sq. metre and thoroughly trowelled in before the paving sets.

BLACK FINISHES OR TILE SLAB

Floor tiles made of PVC

J68: PVC tiles with a minimum thickness of 1.6mm must come from an approved manufacturer.

Fixing must be done using an approved adhesive and following the manufacturer's instructions.

Floor tiles and clay-faced slips

Clay facing slips and clay block floor tiles must come from an approved manufacturer, according to J69. They'll be bedded and joined in 1:3 cement mortar, then pointed in tinted cement on a match.

J70 Dividing Strips: The dividing strips will be 6mm thick ebonite strips cut to length. They must be bedded in the beds and the finished floor level must be flush with it. Between different floor finishes, bay perimeters, and skirtings, dividing strips should be provided.

J71: In this document, dividing strips are not measured separately. The price for pavings should include the cost of their installation wherever it is required.

Orientation

J72: Where the dimensions of the ceilings, walls, or floors are not an exact multiple of the title, the slab or block dimensions used for the ceilings, walls, or floors shall be cut to provide equal margins on both or all sides of the rooms, as the case may be.

Precast Terrazzo

J73: A top finish of at least 19mm thick shall be cast in conjunction with a cement and sand backing on precast terrazzo tiles. The terrazzo finish and backing must be mixed together and then moulded in steel moulds at a pressure of 3.5 N/mm2 (500 ib/sq in) before being used. It must be made up of properly graded aggregate, just like in-situ terrazzo.

J74: The surface must be rubbed down and polished by machine with stones after hardening.

J75: The rates shall include the Architect's/free S.O's colour selection. Precast terrazzo tiles shall be bedded in cement mortar (1:3) and pointed in coloured cement.

Precast Concrete Slabs

J76: Precast concrete slabs shall be made of concrete Grade 20 as described in the concrete work section and in accordance with B.S. 368, natural colour, laid on sand, and joined with cement mortar (1:4) The surface finish will be textured unless otherwise specified.

Laying of Tiles

J77: Tiles must be solidly bedded and laid, with symmetry in relation to the centerline of the floor fixtures and other equipment.

J78: Tiles must be laid with continuous joints and grouted with ordinary or coloured cement slurry unless otherwise specified.

J79: Tiles must be precisely cut and installed on doors, thresholds, walls, openings, projections, and other surfaces.

J80: The completed tile work must be perfectly true and level, or to even falls as required.

J81: After finishing the tile work, soak it in clean water for a few minutes before using it.

J82: Tiles must be approved by the Architect/S.O. in terms of colour, type, and quality.

J83: Terrazzo and marble tiles shall be machine ground with carborundun stone and polished to the Architect's/satisfaction. S.O.'s

J84: For at least 7 days after initial setting, the floor tiles and mortar bed must be continuously wetted with a sand trap or other approved method.

J85: Tiles must be free of flaws, have squared edges, and be brought to the hairline when laid side by side. Safeguarding

J86: Until the project is completed, all work must be adequately covered and protected.

J87: To the satisfaction of the Architect/SO, the entire flooring must be cleaned and left in a sound and perfect condition.

Bills of Quantities Rates

J88: The rates for all in-situ finishes, including beds and backings, shall include the following as well as all making good.

1. Applying to any type of base, including base preparation and any dubbing out that may be required.

2. Concentrate your efforts in small, isolated areas.

3. Work on all edges, ends, angles, arises, and the like, as well as dishing around gullies and outlets.

4. Creating grooves and the like by cutting or forming them

5. Work at a variety of heights

6. Additional labour, such as building frames and the like

7. As needed, provide dividing strips.

8. Protection.

J89: The rates for all sheet finishings and all making good shall include the following ands.

1. Using any type of base and preparing it

2. Creating patterns and working at various heights.

3. Concentrate on small, isolated panels or bays.

4. All types of cutting and fittings, as well as round pipes, tubes, bars, cables, conduits, and the like.

5. Safety and security

SECTION J GLAZING

K1: The glass must be of the types and qualities specified, and must meet all relevant requirements of B.S 952.

K2: All glass must be free of bubbies, smoke waves, air holes, specks, scratches, and other defects, and must be true and even, as well as representative of the specified quality.

K3: Unless otherwise specified, all clear and obscured glass must be 4mm thick and of the highest quality for glazing purposes.

K4: For glazing purposes, plate glass should be approximately 6mm thick, float type, and of the highest quality.

K5: Louvre blades must be made of clear sheet glass that is 5mm thick and has rounded and polished edges.

K6 Putty for metal window glazing shall be a quick-setting tropical putty designed specifically for use with metal windows.

Fixing and Cutting

K7: Within a tolerance of 2mm per 3mm of thickness, all glass shall be cut accurately and fitted to the intended position as shown on drawings.

K8: Edge clearance must be consistent all the way around each pane, with no more than 2 mm ply, and puttied panes must be clipped to metal frames.

K9: The glass must be tight and true, and it must be fixed in such a way that it does not rattle. Angles for glazing, clips, heads, and stops must be precisely set and firmly fixed.

K10: All glass must be bedded in putty or glazing compound according to the manufacturer's instructions, and all corners must be carefully made.

K11: Before puttying, glazing rebates must be printed, and the putty must be finished with a neat fillet and made completely watertight.

K12: Any putty or mastic that fails to set properly or dry out must be finished with a neat fillet and made completely watertight.

K13: When finished, the glazing must be wind and water tight.

Organizing

K14: Smears, excess compound, and sealant must be removed, and cracked or poorly applied putty must be replaced.

K15: Both sides of glass and glazing, as well as metal surfaces, must be cleaned with approved cleaner and left scratch-free.

K16: Clean both sides of all glass thoroughly and replace any cracked or defective glass as directed.

Protection

K17: Ensure that all glazed work is protected from subsequent trades.

Bills of Quantities Rates

K18: Glazing rates must include:

1. Priming, rebate sealing, and other recommended services.

2. Springs, clips, distance piece, and other similar items

3. In wired or pattern glass work, aligning adjacent panes

4. Protection

SECTION K PAINTING AND DECORATING

General

L1: The material and workmanship standards shall not be less than those recommended in the current BS CP 231 'Painting of Buildings' and the relevant British Standards referred to in these specifications.

L2; All paints must be ready-to-use synthetic paints that comply with the general clauses of Section 2 of BS 929.

L3: All paints must be delivered to the job site in their original drums and tins. There will be no dilution of the paint allowed unless it is done strictly according to the manufacturer's instructions.

L4: Undercoat paint must be purchased from the same manufacturer as the finishing coat paint and must be the manufacturer's recommended undercoat for that finish.

L5: To comply with B.S. 1336, the knotting must be an approved patent knotting or shellac.

L6: Stopping is to be done with 'Polyfilla' or something similar that has been approved and used strictly according to the manufacturer's instructions.

L7: Primer paints must be approved patent primers of the appropriate type for the material to be painted, all in accordance with B.S. 2521 – 4. For non-ferrous metals, a zinc chromate primer must be used, and for woodwork, an aluminium primer must be used.

L8: If the timber is to be left with a clear finish, apply one coat of ICI sanding sealer at the time of delivery to the job site.

L9: Painting shall not be done in the rain, fog, or when condensation is likely. No new coat of paint should be applied until the previous one has completely dried. The minimum time between coats must be 16 hours, with a longer time allowed in adhesive drying conditions or when slow hardening materials are used.

L10: During the painting process, provide the necessary scaffolding trestles and protection for the structures.

Surface Preparation is a term used to describe the process of preparing a surface

L11: Rendered surfaces must be sanded smooth and cleaned to the Architect/satisfaction. SO's Minor flaws, cracks, and holes should be repaired and rubbed flush with the surrounding surface after being cut out as needed.

L12: Efflorescence should be brushed off as soon as it appears, and all decoration should be postponed until the efflorescence has faded. Before painting, any moulds or other fungi growth must be treated with an approved fungicidal liquid.

L13: All bar iron and steelwork, including sheeting and pipe, must be thoroughly prepared to the satisfaction of the Architect/S.O. by removing all dirt, rust, grease, and loose mill-scale. Certain meal surfaces may require a chemical treatment with an approved cleaning compound applied according to the manufacturer's instructions.

L14: After the first coat, putty all necessary nail holes and cracks with a putty colour that matches the finish. Before applying a primer coat, make sure that the putty or other approved patching is flush with the adjoining surfaces and that it is sanded smooth and sealed after it has dried.

L15: Before beginning the preliminary painting process, all metal fittings or fastening devices, sockets, outlets, and switches must be removed, cleaned, and reinstalled after the painting work is completed.

Craftsmanship

L16: Prior to priming or decorating, all surfaces must be dry and prepared to the satisfaction of the Architect/S.O.

L17: All surfaces must be primed with the appropriate primer for the surface type.

L18: Any work that is to be fixed in an inaccessible location must be given the full number of coats before being fixed.

L19: Paints must be thoroughly mixed before using them in their original containers.

L20: Thinners should only be used when absolutely necessary and as directed by the manufacturer. Paint must not be tampered with in any way, and different types or brands of paint must not be mixed.

L21: Non-paintable surfaces, such as switches, ironmongery, and so on, must be masked or removed and replaced after the paint has dried.

L22: Coatings must be evenly applied and free of runs, sags, skins, grit, and bristles.

Droppings must be kept off the floor.

L23: Painted woodwork must be knotted, primed, and stopped in the workshop before being delivered to the job site as soon as possible and stored out of the elements. The full number of coats specified must be applied to all top, bottom, and side edges of joinery.

L24: When metal is delivered to a job site primed, it must be thoroughly inspected, with any unsatisfactory priming removed and the surfaces primed with primer.

After preparation, any damaged priming should be touched up. After the metal work has been erected and defective areas have been prepared and printed, a final inspection of the priming will be performed.

L25: The painting shall be done in a good and workmanlike manner, with the Architect/S.O. selecting the tints. Before applying the next coat, each undercoat must be rubbed down with glass paper and thoroughly cleaned. Before applying the next coat, the previous coat must be dry and hard. No paint shall be applied to any surface that contains even the tiniest amount of moisture.

L26: When finished, clean up all drips, splashes, and overpaints around the edges. Make-up and touch-ups are required. Remove all dirt and finger prints.

Panels of Experts

L27: The Contractor shall prepare sample panels for inspection, and no painting shall take place until the samples have been approved.

The Use of Painting Materials

L28: Before using for a different type of material, all brushes, paint rollers, spraying equipment, kettles, and other tools used in the work must be thoroughly cleaned.

L29: When liquid paint containers are opened and before being transferred to paint kettles, all liquid paints must be thoroughly stirred to a uniform consistency.

L30: Before thinning, paste paints must be thoroughly beaten up as directed by the manufacturer.

L31: Oil paint shall be thinned with up to 5% white spirit by volume to maintain a working consistency, except in exceptional circumstances when agreed upon.

L32: The first coat of PVA emulsion paint should be thinned with clear water according to the porosity of the surface to be painted, but not more than 50% by volume. The following coats should not be thinned.

L33: Water paints and distempers must be thinned with clean water to a proper working consistency and strained to remove lumps before use.

L34: Priming coats should be applied with a brush to ensure a uniform thickness and to satisfy the porosity of the surface. Priming should be thoroughly applied to surface joints, angles, and other areas where moisture is likely to collect.

The same day that the steelwork surfaces are cleaned, they must be primed.

L35: Off-site priming coats that have been exposed on the job site or in transit must be touched up or reprimed as needed before undercoating. The Architect/SO may direct repriming if there is any doubt about the primer's ability to fully satisfy the porosity of the surface.

L36: Zinc-based epoxy rein paint as a primer and polyurethane-based epoxy resin paint as a top coat for structural steelwork exposed to weather or aggressive environments.

Woodwork should be painted

L37: After knotting, stopping, and priming, all woodwork must be painted with one coat of undercoat and two coats of gloss finishing coats.

L38: As the work progresses, any paint damage areas must be repaired to the full depth of the damage and properly feathered to the satisfaction of the Architect/S.O.

Woodwork Refinishing

L39: All varnished woodwork must be stopped with stopping to match the wood's colour and rubbed down with fine sandpaper to achieve an even silky finish.

L40: Apply four coats of clear gloss vanish to external woodwork and three coats to internal woodwork.

L41 - Before applying the next coat, each coat should be allowed to dry and lightly rubbed down with fine sandpaper.

Bills of Quantities Rates

L42: The painting and decorating rates must include the following:-

1. Painting the edges of doors and windows that open.

2. Surface preparation and prep work in between coats

3. The removal of door and window furniture, switch covers, and other similar items before beginning preparatory processes and for reinstallation after completion.

4. Use a variety of colours and cut in the edges of flush surfaces.

5. The Architect/S.O. can choose any stock colour he wants and can specify any paint manufacturer's products.

6. Provision of scaffolds, trestles, and other similar items

7. Safety and security

SECTION L ELECTRICAL WORKS

Electrical Work's Scope

M1: The electrical works shall consist of the supply, delivery, and erection of all plant equipment, fittings, and accessories, as well as final testing for the installation of electrical services in accordance with these specifications, schedules, and drawings, as well as the regulations of the supply authority.

M2: Prior to beginning work, all electrical drawings must be read in conjunction with the most recent architectural structural and service drawings.

M3: When installing electrical equipment, special attention should be paid to areas where it must be placed in relation to working tables, wall units, cabinets, patterned walls or ceilings, kitchen areas, and so on.

M4: As applicable, the Contractor shall form chases, holes, and ducts in walls, ceilings, floor slabs, beams, and joists, and make good all work affected. The cost will be considered part of the rates.

M5: The use of British Standards, Codes of Practice, and other standards in this document is intended to clearly and concisely indicate the nature and quality of the items required. If the Contractor wishes to use materials, equipment, or practises that are not covered by these standards, the Architect/S.O must give written approval before any work or orders are placed. The Contractor must describe the alternative along with the reasons for the change in offer.

M6: All relevant aspects of the latest edition of IEE wiring regulations and electrical installation regulations in public buildings must be satisfied and adhered to by the electrical installation.

Plant and Machinery

M7: Before putting any plant or apparatus into operation, the Contractor must ensure that all connections between the plant and the apparatus that may exist or be supplied under the Contract are correct.

M8: Before offering the installation for handover to the Client, the Contractor must include in his tender all work required to bring the plant fully operational, all in accordance with the details contained herein.

M9: The Contractor shall provide all necessary test equipment for conducting running tests on the installations and demonstrating that the equipment meets the design conditions and specifications as installed. Any visits by manufacturers of the equipment to be supplied must be included in the tender.

M10: Materials or items specified by a proprietary manufacturer may be substituted with materials from a different manufacturer, provided that the substitutes are identical to the original specification in every way and that all substitutions are approved in writing by the Architect/S.O. The Architect/S.O will consider all characteristics of the materials or item, such as quality, method of application or fixing, weight, colour, shape, association or connection with other items, and supply evidence of all characteristics when assessing any proposed substitution.

Substitution will not be permitted unless it can be demonstrated that the substituted item will be equal to the specified item in every way.

M11: The Contractor shall instruct the client's staff in the use and proper operation of the works, including emergency procedures, at times agreed upon by the Architect/SO and the Client. The Contractor must be satisfied that such personnel are capable of taking over the installation once it is completed.

M12: Regardless of whether the Architect/S.O. has named or approved suppliers, it is the Contractor's responsibility to ensure that all materials and components meet specifications in terms of manufacture, finish, and performance.

M13: To ensure coordination with other trades and services, the Contractor will communicate with nominated specialist firms working on the site. The terms and conditions set forth by specialist firms should be paid special attention to. If these terms and conditions differ from the terms and conditions set forth herein, the Contractor is responsible for any condition that is not met.

Diagrams of Wiring

M15: For all electrical equipment and/or systems that are part of the Works, the Contractor must provide writing diagrams. Such wiring diagrams must be accurate, complete, and co-ordinate the manufacturer's data for individual pieces of equipment in order to present, as a whole, all of the information for any interconnected group of such items. The Contractor is responsible for the accuracy of such diagrams and is liable for any costs incurred as a result of any errors.

M16: The Contractor shall use labour of sufficient quality to complete the work to the high standard specified in the documents.

Drawings of 'As Installed'

M17: Record drawings and documents, as well as 'As Installed' drawings and documents, must be provided in accordance with the information contained herein.

Material Requirements

M18: All materials used in electrical work must comply with the National Electric Power Authority (NEPA) specification in its most recent edition and amendments. However, if equipment or electrical materials complying with other national standards are desired, they must be approved by the Architect/S.O. in advance.

M19: Protective finishes must be applied to all materials and equipment used in electrical work to ensure that local climatic conditions do not cause deterioration.

M20: Adequate storage and protection of materials on site will be provided as needed to ensure that all items are handed over in complete working order, with all protective finishes intact.

M21: All distribution boards and control units' interior fittings must be painted with a fungus-resistant varnish.

M22: To keep dust and vermin out, all unused holes in distribution boards and similar equipment must be plugged. Where ventilation is required, the holes must be covered with fine mesh wire screens to prevent insects and vermin from entering.

M23: Unless the phrase "or equal" appears, the specialist installer may substitute an item of equivalent technical quality, price, workmanship, and appearance after approval by the Architect/SO.

M24: Samples of such items must be submitted to the Architect/S.O for approval prior to installation. The specialist installer's responsibility to complete the work using approved materials within the contract period is unaffected by the rejection of any sample submitted.

M25: Where proprietary goods, materials, or processes are specified to be used, such goods, materials, or processes must be fixed, used, or carried out in strict accordance with the suppliers' instructions, which must be treated as if they were part of this specification.

Repairing

M26: Builder's work includes forming chases, holes, and ducts in walls, ceilings, floor slabs, beams, joists, and other allied electrical services, as well as trench laying of pipe ducts for electrical and telephone works. These expenses will be factored into the rates.

M27: All apparatus fittings and accessories must be installed with the full number of correct size fixing bolts, screws, and other fasteners specified by the apparatus's manufacturer and appropriate to the surfaces to which they are fastened.

M28: Where plugs are required for fixing equipment to increase block or concrete walls, proprietary plastic plugs must be used. All plug holes must be rotary drilled.

M29: Only cadmium-plated steel screws may be used to secure these plugs; no brass screws are permitted.

M30: Galvanised bolt projecting – type anchor bolts of Messrs Hilti or similar shall be used for heavier fixings into brickwork or concrete, and these shall be grouted into the concrete or blockwork as necessary.

M31: For floor-mounted equipment, galvanised ragbolts must be used. All loose nuts, bolts, washers, and other installation hardware must be galvanised.

Getting to the Top

M32: All accessories and fittings shall be fixed at the following heights from the finished floor level, unless otherwise specified on the drawings.

1. Boards of distribution To the board's centre line, it's 1.9 metres. This height applies to switch fuses and call bells, which must be 2200mm above finished floor level.

2. Lighting fixtures are 2.3 metres long, and switches are 1.370 metres long.

3. Sockets (general) 300mm above the finished floor or work surface

4. Sockets in the workshop of maintenance The height is 1.370 metres.

5. Mount all 13A switch socket outlets at a height of 1400mm above the finished floor level.

6. At a mountain height of 1400mm above finished floor level, all 15A (A/C) switch socket outlets must be installed.

M33: Distribution boards, switchgear, and mechanical plant must be labelled with the area and services fed from them, as well as the source of supply if not otherwise obvious. The engraving must be in white 5mm high letters on black trafolite sheet or equivalent, and it must be screwed or riveted to the lids of the apparatus.

M34: All control switches, insulators, starters, and other similar devices must be labelled with the items or apparatus controlled, as well as the supply voltage and phase.

M35: Where socket outlets and/or single phase isolation in a single room or area are connected to more than one phase, all such outlets and isolators shall be labelled to indicate the phase to which they are connected, and a warning label shall be provided and fixed as directed to indicate the presence of 380 volts between outlets on different phases, as required by the Architect/S.O.

Circuit Diagrams

M36: Each distribution board must be equipped with complete circuit lists, which must be typed on stiff cards and stored in transparent compartments within the board's covers.

M37: The circuit reference, current rating, and service must all be indicated on the lists.

Finishing and Painting

M39: All equipment and materials supplied must be painted or galvanised in general.

M40: Architect/S.O. instructions require that metal be printed with the appropriate primer and painted with three coats of gloss paint.

M41: All damaged paintwork must be repainted to match during transportation or erection.

Clearance, cleaning, and testing of the site

M42: Any excavation required for structures or pole erection must be approved by the Architect/SO, and any surplus soil must be disposed of and the site left tidy.

M43: Before commissioning, all debris and dirt accumulated during installation must be removed, and all electrical equipment must be thoroughly cleaned.

M44: As requested by the Architect/SO or his representative, all equipment must be inspected and tested on site. All tests must be performed at the expense of the contractor, in the presence of, and to the satisfaction of, the Architect/SO, and at such times as he may require.

M45: The Contractor shall serve all notices of testing on the supply authority, pay all fees associated with the testing, and pay any additional charges for retesting to the contractor.

M46: The specialist installer shall conduct tests in accordance with the applicable standard, as well as any additional tests deemed necessary by the Architect/SO, in order to determine that the equipment complies with that standard.

M47: Each item of equipment or installation must have its test results recorded and submitted to the Architect/SO. All test certificates must be forwarded to the Architect/SO in three copies for his records. The specialist installer's obligation is not absolved by the submission of test certificates.

M48: No installation will be accepted or a certificate of completion will be issued until the Architect/SO has approved the tests.

M49: The specialist installer is responsible for providing all testing equipment and instruments.

M50: The results of the tests described below must be submitted to the Architect/SO.

1 When the entire installation is completed in accordance with NEPA regulations, insulation resistance tests are performed on each and every phase.

2. Conduct insulation loop impedance and earth electrode resistance tests as required by NEPA regulations.

3. Proper phase rotation of the entire system.

4. The lighting and small power installation includes polarity of switches and switch socket outlets.

M51: The specialist installer must provide the Architect/SO within two months of the Works being completed.

1. Installation information records and diagrams, and 2. Two copies of manufacturer's maintenance operating instructions, including drawings diagrams and spare parts recommendations for all equipment supplied under the contract.

M52: The specialist installer will be required to keep a set of prints on site throughout the job to keep up-to-date details of the information from which his final record drawings will be prepared. These drawings must be available for inspection by the Architect/SO at any time, but especially before any conduits or other items are hidden.

Installation of a Conduit

M53: Conduit shall be of the highest quality galvanised tubing or heavy gauge plastic, as required by BS standards. They must be at least 20mm in diameter and have an even bore or be free of defects and flaws.

M54: Adequate conduit boxes, junction boxes, covers, and other accessories must be provided to allow for easy cable placement. Bends must be made with the appropriate size bending springs for the conduit diameter.

M55: Appropriate temporary plugging to prevent the ingress of plastic or concrete during construction shall be supplied and fixed for flush installed conduit boxes. This wiring will be left in the outlet boxes for as long as it is necessary and practical.

M56: Concealed conduits in beams and slabs must be cast in such a way that they are precisely positioned in the middle third of the beam or slab. To prevent liquid cement from leaking into the conduit system, joints must be properly sealed.

Prior to beginning work, the Architect/SO must agree on the location of the conduits to be cast in.

M57: Where conduits cross building expansion joints, expansion couplers must be installed at right angles across the joint. Between the two conduit boxes on either side of the expansion joint, a 4 sq.m earth wire must be installed.

M58: The conduit system must be electrically and mechanically continuous throughout the installation, providing an uninterrupted enclosure for the wire cables, particularly where connections are made between the conduit system and the enclosures of other equipment such as switches, outlets, switches, outlets, switchgear, turning, and so on.

Installation of Cables

M59: Copper conductor cables must be drawn into conduits. Single-core non-armoured PVC insulated non-armoured PVC insulated non-armoured non-armoured non-armoured non-armoured non-armoured non

M60: All wiring must be done from point to point; no jointing is allowed.

M61: Wiring will be done using a "loop-in" system and a "T" or other joint. A circuit's live and neutral conductors must always be drawn into the same conduit. Lighting, power, cooking, general services, and so on should all be run through separate conduits.

M62: Separate distribution boards and consumer units' cables must be designed to allow for easy drawing in and out, as well as future extension.

M63: The number and size of cables drawn into a conduit must allow for easy in-and-out as well as future extensions.

M64: The number of sizes of cables drawn into a conduit must be sufficient to allow for easy in-and-out as well as future extensions.

M65: Throughout the installation, all cable runs between one terminal and another must be installed without intermediate joints.

Underground Cables M66: Where underground cables are required to enter a building, these must be contained in ducts. These ducts will be installed with wooden plugs or appropriate size to prevent soil ingress, roding and cleaning of the ducts will be performed after removing the plugs, and the ends of the ducts will be sealed with a mastic compound after drawing cables through the ducts.

M67: Underground ducting must be installed correctly in accordance with requirements and purposes, and close coordination with NEPA and communication authorities is required to ensure adequate information about easy bends, run directions, and other issues.

M68: The incoming ducts shall be 200mm diameter concrete with a 60mm below ground level run and a 1 metre radius bend within the buildings.

Cabling Termination

M69: Armoured cables must be terminated with glands that are appropriate for the job. Where necessary, the glands will be drilled and tapped for installation, and the glands will be filled with PVC shrouds filled with waterproof plastic compound and secured to the cable with PVC tape.

M70: The contractor is responsible for all cable termination work, including providing weatherproof plastic compound PVC shrouds filled with waterproof plastic compound and taped to the cable.

M71: Copper conductors must be used in all cables unless otherwise specified.

M72: Any exposed conductor shall be taped with PVC tapes to the thickness of the original insulation when lugs are soldered to cable ends, the taping being taken partly over the barrel of the cable lug. These tapes must be the same colour as the original insulation.

M73: Only the Architect/direction SO's shall be followed, and all such cable joints shall be made strictly according to the cable manufacturer's recommendations for the cast-resin system, using only materials approved by them.

M74: The strands of cable ends must be well twisted together when used with pinched screw type terminals to make a conductor as solid as possible. Pinched screw type terminals should not be used with single strand cables.

Underground Cable Installation

M75: With the drum supported on a spindle and cable drum jacks, cables must be removed from drums. Cables must be fed into horizontal routes from the underside of the drum, and cables must be fed into manholes from the top of the drum.

M76: Care must be taken to break the drum's rotation and prevent the drum's cable turns from becoming loose.

M77: Cables must not be dragged across the ground, concrete, or other surfaces; instead, they must be properly supported on a roller and/or manhandled into place.

M78: In the bottom of the trench, the specialist installer must lay an 8 cm layer of finely sifted soil or sand on top of which the cable will be laid. On top of this layer, an additional 8cm layer of finely sifted soil or sand must be laid, followed by approved type interlocking cable over tiles.

Boards of Distribution

M79: All miniature circuit-breaker distribution boards must be approved, metal clad, and surface mounted, with all parts rust-proof and stove enamelled.

M80: The distribution boards must be supplied and installed in the positions indicated on the drawings by the specialist installer.

M81: All distribution boards must be flush with the surface and/or mounted on the floor, and they must include an integral insulator.

M82: The distribution boards must include all gland plates, cable entries, fixing brackets, and supports that are appropriate for the cable specified and the location indicated.

M83: Molded plastic cover strips must be installed on unused spare switch units. The panel must be equipped with all necessary components for the installation of future circuit breakers. Breakers should be single pole for single phase line to neutral and triple pole for three phase, depending on the final sub-circuit.

M84: Wiring shall be 1000 V Grade PVC insulated cables with copper conductors drawn into surface mounted heavy gauge steel conduit, surface mounted steel trunking, and concealed plastic (PVC) conduit as applicable and described.

Miniature Circuit Breakers

M85: The miniature circuit breaker (M.C.B) must be made by a reputable company.

M86: The M.C.Bs must adhere to the following guidelines.

1. On triple pole miniature circuit breakers, each pole shall be a separate tripping mechanism, the toggle assemblies of all three poles shall be internally mechanically linked for simultaneous isolation of all three poles under fault conditions, and each pole's overload tripping characteristics and calibrations shall be completely unaffected by the loading of its neipole.

2. Temperature fluctuations and high ambient temperatures should have no effect on the time delay tripping mechanism and calibration. The M.C.Bs must detect and isolate a sustained load of 35 percent above the rated load at 35 degrees C, rather than trip through the overload protection at full load.

3. High rupturing capacity (HRC) fuse links with base shall be used on all circuit breakers rated between 100 and 500 amps. Circuit breakers over 500 amps must be magnetic.

Circuit Breakers for Earth Leakage

M87: All Earth Leakage Circuit Breakers (ELCB) must be made by a manufacturer that has been approved.

Earthing is a term that refers to the

General

M88: The earthing system must be installed in accordance with NEPA regulations.

M89: The specialist installer is responsible for supplying, installing, and connecting all equipment required for a proper earthing system.

Earth Continuity

M90: The entire electrical installation, including all cable armouring or metallic sheathing, must be electrically continuous throughout, forming a fully bonded earth system.

All apparatus and parts thereof that are not solidly connected to it by copper conductors in an approved manner must be secured with substantial bonding clamps.

M91: All switched socket outlets' third pin, all lamp holders' metallic parts, and all switch boxes must be effectively earthed.

M92: All conduit boxes must be drilled and tapped to a minimum M6 size for earth connection with a cadmium plated terminal type washer or equivalent, and secured with a cadmium plated brass screw.

M93: The specialist installer must ensure that every complete earth loop circuit, including conduits, cable sheaths, core conductors, and other components, has an impedance value that does not exceed that specified in the most recent edition of the NEPA Regulations.

Connection to a Switch

M94: Each single-pole switch must be installed only on the line. It must be a linked switch if a switch is connected to the earthed conductor.

Electric Motors

M95: A starting and stopping device must be installed on every electric motor. The devices must be placed in such a way that the person in charge of the motor can easily operate them.

M96: Overload and time lag devices must be installed on all motors. Independent or group isolating switches must also be installed on the motors and starters.

M97: When mechanically coupling a motor to the load it drives, the axes of the shafts for direct coupling and/or the pulleys for belt drives must be aligned.

M98: Motors with more than 12 horsepower must have starters with under-voltage protection and starter current limiting devices.

M99: Separate switchfuses must be used for lift and elevator motors. The circuits must be clearly labelled with the words "Lift" or "Elevator." The conductors feeding the lift motor, the control circuits, and any low voltage signalling device must all have their own trailing cables.

Temporary Installation

M100: A temporary installation is one that is used to supplement a permanent electrical service.

Temporary installations, on the other hand, must be completely overhauled at least once every three months.

M101: Temporary installations must be designed to the same standard as permanent installations to avoid hazards on the construction site. Only cables may be fixed for a limited time if they are protected from mechanical damage and dampness and are not likely to be touched by untrained individuals.

M102: On temporary installations, all switches must be double-pole.

M103: Wherever there are obstacles, pitfalls, or other hazards, adequate lighting must be provided.

M104: All portable lamps must have an insulated headlamp and an insulated guard.

Precautions against Fire and Explosion

M105: Only high-temperature cables should be used in ducts or areas where ambient temperatures are above the following levels (B.S)

Type of Silicon Mineral Impregnated

Insulation Rubber PVC Paper Rubber Insulation

Maximum duct or area temperature in degrees Celsius (oC) 55 65 75 145 145 M106: Special heat-resistant cables must be used if the temperature exceeds 145 degrees Celsius.

M107: In boiler houses and similar locations where cables are connected to thermostats, immersion heaters, and other equipment installed near or on the boiler, the main wiring can be done with standard cables in conduits. To allow easy removal of the thermostat or other apparatus, any connection adjacent to the boiler must be made with heat-resistant cable enclosed in flexible metallic conduit.

M108: Pre-insulated ables, preferably on porcelain cleats, must be used in areas exposed to acids or alkalis. Avoid using metal as a covering. Steel conduits and other ferrous metal-based systems should also be avoided near seawater.

M109: Conductors that are exposed to flammable environments must be built or protected using methods that adequately prevent danger to people or property.

M110: When installing conductors in a building, all necessary precautions must be taken to ensure that there are no gaps or holes in the walls or floors that could aid in fire spread.

M111: No apparatus shall be installed in places exposed to water, oil, steam, vapour, or any other form of mechanical damage unless properly protected.

Installation of Telephone Sockets

M112: The contractor is responsible for providing and installing a complete conduit system to facilitate the installation of the telephone system as shown on the drawings. The conduits must be constructed as previously described, with adequate draw-in boxes and draw-in wires.

M113: Except where otherwise specified, the wiring, supply, installation, and commissioning of telephone/intercom equipment shall be a separate contract to be completed by others.

Alarm System for Fires

M114: Both a manual and an automatic fire alarm system are required. Detectors, a control panel/standby battery-plus-charger (24V d.c), and fire alarm bells are all required.

M115: Break-glass type manual detectors, Thorn EMI Protech or approved equivalent are required.

M116: Automatic smoke detectors must be of the ionisation type. Thorn EMI Protech or an approved substitute is required.

M117: Fire alarm bells must be dome-shaped, 150 mm in diameter, Thorn EMI Protech or approved equivalent.

M118: The fire alarm control panel must include a battery and charger, as well as supervisory and visual indicators to monitor all of the building's alarm initiating points.

Prices are all-inclusive.

M119: The Contractor shall include all items requested under the preceding clauses in the rates he inserts in the Bills of Quantities.

Bills of Quantities Rates

M120: All pipework and ductwork rates must include the following:

1. Running length joints and couplers, as well as standard pipe supports (such as clips saddles, pipe hooks, holderbats, and brackets).

2. Connecting to fittings and other components

3. Sockets, short lengths, and short running lengths

4. Brackets, hangers, and other supports are cut and pinned in, plugged and screwed or nailed in.

5. Forming or cutting all holes, cutting away, and making good in general.

6. Where pipes pass through walls or floors, etc., pipe sleeves are provided.

7. Testing and protection

EXTRA ELECTRICAL WORKS

ULBS AND FLUORECENTS

L1 BULBS

Description: 1 x 60 watt,

wall bracket fitting c/w tube

Size: 1 x 60w

Number 10 x 3

L2 BULBS

Description: 2 x 50w ceiling fitting, verandah

Size: 2 x 50w

Number: 11 x 3

L3, L5, L6, L7, L8 BULBS

Description: 1 x 40w Surface mounted slim section fluorescent batten fitting with three position rotating end-caps.

Size: 1x 40w

Number: 40 x 3

L4 BULBS

Description: 1 x 60w downlight screw neck, bulk head fitting, bath and toilet rooms

Size: 1 x 60w

Number: 9 x 3

CEILING FANS AND CONTROL SWITCH Ceiling fans

L3, L5, L6, L7, and L8

Description: 65W Newcline ceiling fan

Size: 65W

Number: 20 x 3

CONTROL SWITCH

Description: Ceiling fan's Control Switch

Size:

Number: 20 x 3

SWITCH AND OUTLET SOCKETS Switch 1 gang, l way switch

Description: 10A, 1gang, 1way switch model large rocket switch legrand; CAT.NO.614321

Size: 10A

Number: 7 x 3

2 gang, 2 way switches,

Description: 10A, 2gang, 2way switch model large rocket switch legrand; CAT.NO.614322

Size: 10A

Number: 8 x 3

3 gang, 2 way switches,

Description: 10A, 3-gang, 2way switch model large rocket switch legrand; CAT.NO.614321

Size: 10A

Number: 1 x 3

SOCKET OUTLETS Single switch socket outlets

Description: 13A, single switch socket outlet Neon light mounted at 300mm above finish floor level low legrand model socket CAT.NO.614265

Size: 13A

Number: 11 x 3

Twin switch socket outlets

Description: 13A, Twin switch socket outlet Neon light mounted at 300mm above finish floor level low legrand model socket.

Size: 13A

Number: 6 x 3

CABLES

Material: Copper conductor

Description: Cables drawn into conduits shall be copper conductors, PBC insulated, non armoured single-core 600 volts grade.

Size: 600 Volts Grade

DISTRIBUTION BOARDS

Metal clad miniature circuit breakers (MCB) distribution boards with the number of ways and capacities shown on the relevant drawings are required. As specified, flush or surface mounted, with earthen busbar and hinged covers. Individual moulded cases with magnetic hydraulic time delay mechanism and a 25% spare way made available shall be used as circuit breakers in this distribution board.

ELCB (60A MCCB TPN)

Where earth leakage circuit breakers (ELCBs) are installed for use as a main switch/breaker, they must be voltage operated and of the make, type, and rating specified on the relevant drawings.

SECTION M MECHANICAL WORKS

Water Closet and Cistern

Description: Low level wash down WC with white glazed pan in P-Tray, solid plastic seat, and covered 9litres Cistern with ball valve, flush master.

Size:

Number: 6 x 3

Wash-hand Basin

Description: 560 x 410 x 310 Amitage shank (WHB) wash hand basin, white glazed with 13mm hot-cold crame plated water pillar tap, 32mm bottle trap bracket, chain and plug.

Size:

Number: 3 x3

www.ingramcontent.com/pod-product-compliance
Lightning Source LLC
Chambersburg PA
CBHW060824220526
45466CB00003B/972